Bearing Dynamic Coefficients in Rotordynamics

Wiley-ASME Press Series

Computer Vision for Structural Dynamics and Health Monitoring
Dongming Feng, Maria Q. Feng

Theory of Solid-Propellant Nonsteady Combustion
Vasily B. Novozhilov, Boris V. Novozhilov

Introduction to Plastics Engineering
Vijay K. Stokes

Fundamentals of Heat Engines: Reciprocating and Gas Turbine Internal Combustion Engines
Jamil Ghojel

Offshore Compliant Platforms: Analysis, Design, and Experimental Studies
Srinivasan Chandrasekaran, R. Nagavinothini

Computer Aided Design and Manufacturing
Zhuming Bi, Xiaoqin Wang

Pumps and Compressors
Marc Borremans

Corrosion and Materials in Hydrocarbon Production: A Compendium of Operational and Engineering Aspects
Bijan Kermani and Don Harrop

Design and Analysis of Centrifugal Compressors
Rene Van den Braembussche

Case Studies in Fluid Mechanics with Sensitivities to Governing Variables
M. Kemal Atesmen

The Monte Carlo Ray-Trace Method in Radiation Heat Transfer and Applied Optics
J. Robert Mahan

Dynamics of Particles and Rigid Bodies: A Self-Learning Approach
Mohammed F. Daqaq

Primer on Engineering Standards, Expanded Textbook Edition
Maan H. Jawad and Owen R. Greulich

Engineering Optimization: Applications, Methods and Analysis
R. Russell Rhinehart

Compact Heat Exchangers: Analysis, Design and Optimization using FEM and CFD Approach
C. Ranganayakulu and Kankanhalli N. Seetharamu

Robust Adaptive Control for Fractional-Order Systems with Disturbance and Saturation
Mou Chen, Shuyi Shao, and Peng Shi

Robot Manipulator Redundancy Resolution
Yunong Zhang and Long Jin

Stress in ASME Pressure Vessels, Boilers, and Nuclear Components
Maan H. Jawad

Combined Cooling, Heating, and Power Systems: Modeling, Optimization, and Operation
Yang Shi, Mingxi Liu, and Fang Fang

Applications of Mathematical Heat Transfer and Fluid Flow Models in Engineering and Medicine
Abram S. Dorfman

Bioprocessing Piping and Equipment Design: A Companion Guide for the ASME BPE Standard
William M. (Bill) Huitt

Nonlinear Regression Modeling for Engineering Applications: Modeling, Model Validation, and Enabling Design of Experiments
R. Russell Rhinehart

Geothermal Heat Pump and Heat Engine Systems: Theory and Practice
Andrew D. Chiasson

Fundamentals of Mechanical Vibrations
Liang-Wu Cai

Introduction to Dynamics and Control in Mechanical Engineering Systems
Cho W.S. To

Bearing Dynamic Coefficients in Rotordynamics

Computation Methods and Practical Applications

Łukasz Breńkacz
Institute of Fluid Flow Machinery
Polish Academy of Sciences
Gdańsk, Poland

This Work is a co-publication between John Wiley & Sons Ltd and ASME Press.

Registered Office
John Wiley & Sons, Inc., 111 River Street, Hoboken, NJ 07030, USA

Editorial Office
111 River Street, Hoboken, NJ 07030, USA

For details of our global editorial offices, customer services, and more information about Wiley products visit us at www.wiley.com.

Wiley also publishes its books in a variety of electronic formats and by print-on-demand. Some content that appears in standard print versions of this book may not be available in other formats.

Library of Congress Cataloging-in-Publication Data

Names: Breńkacz, Łukasz, author.
Title: Bearing dynamic coefficients in rotordynamics : computation methods and practical applications / Łukasz Breńkacz.
Description: First edition. | Hoboken, NJ : Wiley, 2021. | Includes bibliographical references and index.
Identifiers: LCCN 2020053700 (print) | LCCN 2020053701 (ebook) | ISBN 9781119759263 (hardback) | ISBN 9781119759249 (adobe pdf) | ISBN 9781119759171 (epub) | ISBN 9781119759287 (obook)
Subjects: LCSH: Rotors–Dynamics. | Bearings (Machinery)
Classification: LCC TJ1058 .B74 2022 (print) | LCC TJ1058 (ebook) | DDC 621.8/2–dc23
LC record available at https://lccn.loc.gov/2020053700
LC ebook record available at https://lccn.loc.gov/2020053701

Cover Design: Wiley
Cover Image: © photosoup/iStock/Getty Images

Set in 9.5/12.5pt STIXTwoText by SPi Global, Pondicherry, India
Printed and bound by CPI Group (UK) Ltd, Croydon, CR0 4YY

C9781119759263_190321

to my wife Dagmara, my daughter Agata, and my son Wojciech

Contents

List of Figures

List of Tables

Preface

This monograph concerns the experimental and numerical methods of determination of dynamic coefficients of hydrodynamic radial bearings. Bearings are one of the basic elements influencing the dynamics of rotor machinery. The main parameters with which the operation of bearings can be described (and thus the operation of the entire rotating system) are their stiffness and damping coefficients.

This book includes a chapter about practical applications of bearing dynamic coefficients. It is shown how changes of bearing dynamic coefficients affect the dynamic performance of rotating machinery. Some examples are included with all the necessary data to allow rotordynamics analysis to be conducted and the dynamic coefficients of journal bearings to be calculated so that the readers can replicate the results presented in this book and compare them with their own results. This book presents in detail an experimental method of determining dynamic coefficients of bearings. An additional objective is to describe numerical methods of determining dynamic coefficients of hydrodynamic bearings (linear and non-linear). The range of applicability of various calculation methods was determined based on measurements made for a rotating machine equipped with hydrodynamic bearings with clearly non-linear operating characteristics.

Experimental research was carried out with the use of the impulse method, on the basis of which dynamic parameters of hydrodynamic bearings were determined. The applied method with a linear calculation algorithm allows the determination of stiffness and damping coefficients and the determination of mass coefficients in one algorithm. The stiffness and damping coefficients cannot be determined directly, thus indirect calculation methods are used. The mass of the rotor is a directly measurable parameter. Indirectly calculated mass coefficients can be compared with the known mass of the rotor. On this basis, it is possible to make preliminary estimations of the correctness of the results obtained.

As part of the study, the sensitivity analysis of the aforementioned experimental method was carried out with the use of a model created in Samcef Rotors software. The influence of unbalance, displacement of measuring sensors, and various variants of driving force were analyzed. Based on experimental research, dynamic coefficients of hydrodynamic bearings in a wide range of rotational speeds, taking into account resonance speeds and higher speeds, were determined. They were verified using Abaqus software.

Numerical calculations of stiffness and damping coefficients of hydrodynamic bearings with the use of linear and non-linear calculation models developed by IMP PAN in Gdańsk were also carried out. The obtained results were verified. The stiffness and damping parameters of hydrodynamic bearings determined using numerical models (linear and non-linear) were compared with the results of experimental research. From this comparison it was possible to evaluate the differences in the values of dynamic coefficients of bearings calculated on the basis of linear and non-linear numerical methods and the experimental method.

I would like to thank all employees of the Turbine Dynamics and Diagnostics Department of the Polish Academy of Sciences for their kindness, for the atmosphere of friendship that they surrounded me with throughout the entire period of work, and for the fact that I could always count on their support and research experience. In particular, I would like to express my gratitude to Professor Grzegorz Żywica and Professor Jan Kiciński.

Gdańsk, November 2020 *Łukasz Breńkacz*

Symbols and Abbreviations

Symbols

$\beta_{i,k}$	dimensionless damping coefficients of lubricating film, $i,k = x,y$
$\gamma_{i,k}$	dimensionless stiffness coefficients of lubricating film, $i,k = x,y$
μ	dynamic viscosity of oil, $N{\cdot}s/m^2 = Pa{\cdot}s$
μ_o	oil viscosity at temperature T_0, $N{\cdot}s/m^2 = Pa{\cdot}s$
Π	dimensionless hydrodynamic pressure, $\Pi = p(\Delta R/R)^2/\mu_0\Omega$
Π^*	pressure at static equilibrium point
σ	standard deviation
τ	dimensionless time $\tau = \omega t$
ψ	angular coordinate $\psi = \frac{x}{R}$, rad, angles defining position of outlet and inlet edges of lubricating pockets
ω	frequency, rad/s, rotational speed, rpm
Ω	journal angular velocity, rad/s

Abbreviations

A	integral values according to Eq. (9.2)
B	integral values according to Eq. (9.2)
c	damping
$c^s{}_{ij}$	damping coefficients of bearing no. s, N·s/m, $s = 1,2$; $i,j = x,y$
d	shaft and bearing journal diameter, m
D	disk diameter, m
Dz	bearing housing diameter, m
$D^s{}_{ij}$	displacement of bearing no. s in the frequency domain, m, $s = 1,2$; $i,j = x,y$
F_{Ni}	force in the N direction, $N = x,y$
$F^s{}_{i,j}$	flexibility vector of bearing no. s, m/N, $s = 1,2$; $i,j = x,y$
F_1, F_2	functions describing dynamic coefficients of bearings
$H^s{}_{i,j}$	stiffness vector of bearing no. s, N/m, $s = 1,2$; $i,j = x,y$
h	lubrication gap thickness, m

H	dimensionless thickness of lubrication gap, $H = h/\Delta R$
k	stiffness
k^s_{ij}	stiffness coefficients of bearing no. s, N/m, $s = 1{,}2$; $i{,}j = x{,}y$
L	bearing housing width, m
m	mass
m^s_{ij}	mass coefficients of bearing no. s, kg, $s = 1{,}2$; $i{,}j = x{,}y$
O_c	center of bearing journal
O_p	center of the coordinate system associated with the bearing housing P_0 force acting on the bearing
p	hydrodynamic pressure in the lubricating film, Pa
R	journal radius, m
ΔR	radial bearing backlash, m
S_0	Sommerfeld number (dimensionless load-bearing capacity of bearings), $S_0 = P_{st}(\Delta R/R)^2/LD\mu_0\Omega$
t	time, s
u, v, w	speed components of the liquid element in the coordinate system x, y, z, m/s
U_1, V_1, W_1	components of lower sliding surface speed, m/s
U_2, V_2, W_2	components of upper sliding surface speed, m/s
W_x, W_y	dynamic components of the reaction of lubricating film, N
ΔW_x, ΔW_y	changes in the dynamic components of the reactions of lubricating film, N
W_0	static components (at equilibrium point) of the reactions of lubricating film, N
$\bar{x}$	mean value
$x{,}y{,}z$	coordinates
$X_c{,}Y_c{,}Z_c$	rectangular coordinate system related to the position of the journal center $(O_c)_{st}$ at the static equilibrium point, m
$X_p{,}Y_p{,}Z_p$	rectangular coordinate system related to the bearing housing center O_p, m
$X_c{,}Y_c{,}Z_c$	dimensionless rectangular coordinate system related to the position of the journal center $(O_c)_{st}$ at the static equilibrium point ($X_c = x_c/\Delta R$)
$\widetilde{X}_c$, $\widetilde{Y}_c$	disturbing parameters
Z	matrix of dynamic coefficients of bearings

About the Companion Website

The book is accompanied by a companion website:

www.wiley.com/go/brenkacz/bearingdynamiccoefficients

The website includes:

- Computational codes
- Recordings

1

Introduction

The analysis of dynamic properties of rotating machinery has for many years been the subject of numerous research studies carried out in many scientific centers. For modern day rotating machinery it is required to work with increasingly difficult operating parameters while maintaining a light and compact design. Increased efficiency, reliability, and precision are also required. Rotating machinery with hydrodynamic bearings is used in many sectors of the economy, e.g. energy, transport, aviation, and military. Very often they are a key element of large technical objects.

In steam turbines used for energy conversion, one of the key components are hydrodynamic plain bearings. These machines are referred to as “critical machinery,” i.e. they are required to be extremely reliable. Unplanned downtime due to poor technical condition leads to significant financial losses. They are therefore monitored and thoroughly analyzed.

The starting point for the analysis of hydrodynamic radial bearings are the equations of motion. From the mathematical perspective, we are dealing with non-linear differential equations (motion equations of the entire rotor) related to the system of partial differential equations describing the properties of plain bearings and the supporting structure. This association occurs through the stiffness and damping coefficients of hydrodynamic bearings. These coefficients determine the dynamic properties of the bearings. Linear and non-linear numerical calculation models are available, but the limit from which the much more complex and time-consuming non-linear models must be used is not clearly defined. An experimental (using a linear algorithm) and numerical (using linear and non-linear algorithms) calculation of dynamic coefficients of bearings for a wide range of rotational speeds was carried out in order to elaborate on the issue formulated in the title of this monograph.

1.1 Current State of Knowledge

The values of stiffness and damping coefficients have a decisive impact on the analysis of rotary machine vibrations. During the dynamic analysis of the rotating shaft, it is necessary to build a discrete model, define the boundary conditions, and calculate the values of these coefficients. The stiffness and damping coefficients change with rotational speed. For the majority of bearings in operation, the coefficients are non-linear in nature,

Bearing Dynamic Coefficients in Rotordynamics: Computation Methods and Practical Applications,
First Edition. Łukasz Breńkacz.

Companion website: www.wiley.com/go/brenkacz/bearingdynamiccoefficients

which means that they change in time and are dependent on the driving force (Kiciński 2006). For most jobs, a linearized form of coefficients is used, which means that they have constant values for a given speed and do not change for different values of driving forces. It should be remembered that dynamic coefficients also change with changes in operating temperature, bearing supply pressure, and bearing load (Hamrock et al. 2004).

For many years at the Institute of Fluid Flow Machinery in Gdańsk under the management of Professor Jan Kiciński, programs from the MESWIR series for numerical calculation of dynamic coefficients of hydrodynamic bearings and rotor dynamics together with imperfections have been developed. A key element in the calculation is the appropriate determination of the stiffness and damping coefficients of hydrodynamic bearings. In the NLDW software (one of the programs of the MESWIR series) it is possible to determine the non-linear form of these coefficients. This is the basis for further dynamic analysis of the entire system, which can be conducted on many planes. Unfortunately, only an indirect comparison of the results of numerical calculations with a real model is possible, i.e. by comparing the amplitudes of vibrations measured on the basis of experimental tests and the amplitudes of vibrations calculated on the basis of numerical analyses. This work presents a method which enables dynamic coefficients of bearings to be determined on the basis of experimental research. They can be directly compared with the results of numerical tests.

Numerical determination of stiffness and damping coefficients of bearings is most often performed by solving the Reynolds differential equation. For linear systems with known parameters such calculations are performed with very high accuracy (Duff and Curreri 1960; Giergiel 1990). For systems for which it is necessary to describe using methods with non-linear calculation algorithms (Hayashi 1964), or methods with complex structure with unknown parameters, calculations become more complicated and are often plagued with significant errors (Fertis 2010). During the calculation it is necessary to take into account the relationships between different parts of the system such as rotor, bearings, and supporting structure (Kiciński and Żywica 2014a; Mikielewicz et al. 2005).

The principles of hydrodynamic lubrication of bearings and the principles of formation of a wedge of lubricant are described in many books on the theory and practice of plain bearings (Neyman and Sikora 1999; Wierzcholski 1994). Numerical linear and non-linear algorithms for calculating stiffness and damping coefficients of hydrodynamic radial bearings are presented in an accessible way in the book *Teoria i badania hydrodynamicznych poprzecznych łożysk ślizgowych* (*Theory and Research on Hydrodynamic Radial Plain Bearings*) (Kiciński 1994). The hydrodynamic bearing can be treated as a mechanical system consisting of elements of varied stiffness. The properties of the bearing are determined by the lubricating film that forms between the journal and the bearing housing (Sikora 2009). Bearing properties are defined as the properties of a lubricating film.

A schematic view of a hydrodynamic radial bearing is shown in Figure 1.1. In the center of the bearing housing there is a rectangular system of coordinates X_p, Y_p with the center marked as O_p. Under conditions of static equilibrium a relationship occurs (1.1), where W_0 is the reaction of the lubricating film to the force P_0, while X_p, Y_p, Y_p determine the position of the center of the journal O_c.

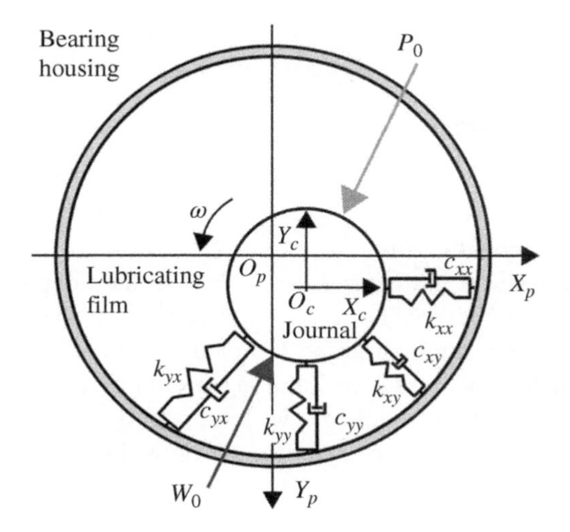

Figure 1.1 Lubrication film model for small journal displacement.

$$W_0 = f\left(x_p, y_p\right) \tag{1.1}$$

It is important to note that the $f(x_p, y_p)$ function, which determines the relationship between the reaction of the film and the displacement of the journal, is non-linear, and this non-linearity increases strongly with the increase of displacements of x_p, y_p. The reaction W_0 can be written as the sum of the reactions W_x and W_y in two perpendicular directions. Definition of the stiffness coefficients of the lubricating film are determined by Eqs. (1.2) and (1.3):

$$k_{xx} \approx \frac{\Delta W_x}{\Delta x} = \frac{\partial W_x}{\partial x}, \qquad k_{xy} \approx \frac{\Delta W_x}{\Delta y} = \frac{\partial W_x}{\partial y} \tag{1.2}$$

$$k_{yy} \approx \frac{\Delta W_y}{\Delta y} = \frac{\partial W_y}{\partial y}, \qquad k_{yx} \approx \frac{\Delta W_y}{\Delta x} = \frac{\partial W_y}{\partial x} \tag{1.3}$$

where ΔW_x, ΔW_y are the changes in the reaction of the film due to a small increase in force ΔP at the point of static equilibrium, and Δx, Δy are the changes in journal displacements due to an increase in force ΔP calculated at the point of static equilibrium. A characteristic feature of radial plain bearings is their inequality of "cross-coupled" reaction and displacement ratios, i.e. inequality of coefficients $k_{xy} \neq k_{yx}$. This is not a property of linear-elastic mechanical systems. It turns out that it is the cause of hydrodynamic instability.

The stiffness of the lubricating film, as an analogy of mechanical systems, often assumes that x_p, y_p displacements are slow. However, in hydrodynamic bearings, when the journal performs small, fast oscillations around the static equilibrium point, significant additional components of the W_x and W_y reaction may arise due to the resistance to motion in the viscous lubricant. In addition to the elastic properties, it is also necessary to determine the damping properties of the lubricating film. They can be formulated using Eqs. (1.4) and (1.5) in the same way as for the stiffness coefficients:

$$c_{xx} = \frac{\partial W_x}{\partial \dot{x}_c}, \quad c_{xy} = \frac{\partial W_x}{\partial \dot{y}_c} \tag{1.4}$$

$$c_{yy} = \frac{\partial W_y}{\partial \dot{x}_c}, \quad c_{yx} = \frac{\partial W_y}{\partial \dot{y}_c} \tag{1.5}$$

where $\dot{x}_c$, $\dot{y}_c$ mean the derivative of the displacement in time.

In numerical calculations, the following relationship is used, i.e. in case of damping there is no anisotropy of "cross-coupled" coefficients, i.e. $c_{xy} = c_{yx}$ (Kiciński 1994). Assuming small displacements x_c, y_c, elastic and damping properties of the lubricating film can be described by four stiffness coefficients and four damping coefficients. In numerical analyses three damping coefficients can be obtained. The components of the reaction of the lubricating film – W_x, W_y – can be represented by linear relationships (1.6) and (1.7):

$$W_x = k_{xx}x_c + k_{xy}y_c + c_{xx}\dot{x}_c + c_{xy}\dot{y}_c \tag{1.6}$$

$$W_y = k_{yx}x_c + k_{yy}y_c + c_{xy}\dot{x}_c + c_{xx}\dot{y}_c \tag{1.7}$$

The dynamic components W_x,W_y depend on the momentary position of the center of the journal x_c, y_c. Assuming that k_{xx}, k_{xy}, k_{yx}, $k_{yy} = k_{i,k}$ and c_{xx}, c_{xy}, c_{yx}, $c_{yy} = c_{i,k}$ these coefficients can be described by Eq. (1.8), where the functions F_1 and F_2 are non-linear. W_x and W_y relationships form the basis for a linear description of dynamic properties of the lubricating film and rotor–bearing systems (Dąbrowski 2013). They can be used to quickly and easily determine a number of very important characteristics of hydrodynamic bearings. Due to the non-linear nature of bearings, a linear description can only reflect the actual bearing properties and the system associated with them to a limited extent. In this monograph the results of calculations using linear algorithms are shown on a specific example of a laboratory test rig operating with parameters which should be described using methods with non-linear algorithms.

$$k_{i,k} = F_1\left(x_p, y_p\right), \; c_{i,k} = F_2\left(x_p, y_p\right) \tag{1.8}$$

The non-linear description is much more complicated than the linear one (Batko et al. 2008; Minorsky 1967; Skup 2010), but it allows (by determining the trajectory of the journal) to analyze e.g. self-excited vibrations of the system or to determine the influence of various types of external forces and damage. It provides much greater opportunities for theoretical analysis of the bearing's operation and the system associated with it. However, its use may be limited by the time of numerical calculations and computing capabilities of computers. As the computing power of computers increases, these limitations become less and less significant.

The methods most frequently used in non-linear analysis consist of the use of non-linear equations of motion and the principle of superposition to solve the Reynolds equation, where at each time point the part of the equation referring to the so-called combined effect of the rotational speed of the journal (the "kinetostatic" part) and to the so-called extrusion effect (the "dynamic" part) is solved separately. The above principle is the basis for the vast majority of numerical methods of determining the trajectory of a journal in the case of large displacement, e.g. methods developed by Booker, Block, and other authors (Blok 1975; Booker 1965). They have one common flaw, i.e. when the solutions are compiled different boundary conditions are used in the "kinetostatic" part and in the "dynamic" part of the Reynolds equation (Kiciński 1994).

It is possible to formulate a non-linear description without the above-mentioned flaw (Kiciński 1994). This description can be based on four predefined stiffness coefficients and four damping coefficients and is used in the MESWIR environment. If the whole interval of variation of the journal trajectory is divided into a sufficiently large number of successive subintervals at small intervals of Δt, it may be assumed that there are areas of the circle of backlash corresponding to those subintervals in which the stiffness and damping coefficients are approximately constant. Each such area will have constant coefficients which need to be determined. The characteristic feature of this method of determining the stiffness and damping coefficients is the fact that for each such area the full Reynolds equation is solved with only one boundary conditions (and thus without the use of superposition solutions). The dynamic components W_x, W_y depend not only on the momentary position of the center of the journal x_c, y_c, but also on the speed of change of this position $\dot{x}_c$, $\dot{y}_c$, according to Eq. (1.9) (Buchacz et al. 2013).

$$k_{i,k} = F_1\left(x_p, y_p, \dot{x}_p, \dot{y}_p\right),\ c_{i,k} = F_2\left(x_p, y_p, \dot{x}_p, \dot{y}_p\right) \tag{1.9}$$

The functions F_1 and F_2 are non-linear, as in the case of small vibrations around the static equilibrium point. Based on the above at each time point of the trajectory t_{k-1}, t_k, t_{k+1}, etc., and based on the determined stiffness and damping coefficients, it is possible to determine the increments of displacement Δx and Δy in a stepwise procedure. The accuracy of the calculation will increase as the time step decreases, i.e. as the areas in which the stiffness and damping coefficients can be assumed to be constant decrease. This relationship will be true until a certain point, and with too small time steps the accuracy of the calculations will be limited. Figure 1.2 illustrates the difference between linear and non-linear analysis based on stiffness and damping coefficients. The hatched fields show the areas where the coefficients are approximately constant.

The numerical analysis of vibrations with regard to large displacements of the journal carried out with the methods discussed above, regardless of whether it uses the principle of superposition or not, which is described as non-linear, is in fact a linear analysis of "pseudostatic" states (Kiciński 1994). In this context, bearing dynamics is a sum of separate "pseudostatic" states, in which time is treated as a parameter. In practice, the state of the bearing at any time t_k is influenced by phenomena occurring at moments preceding the moment under consideration, i.e. t_{k-1}, t_{k-2}, etc. This means that it is necessary to "memorize" the entire history of previous states and a continuous mathematical description in which time is no longer a parameter but is the third (next to geometric coordinates) independent variable.

In the case of a description with "prehistory", the time relationships used ensure that the Reynolds equation will not be integrated in an "empty" lubrication gap. In the MESWIR environment, where numerical analyses were performed, a continuous description with "prehistory" is used.

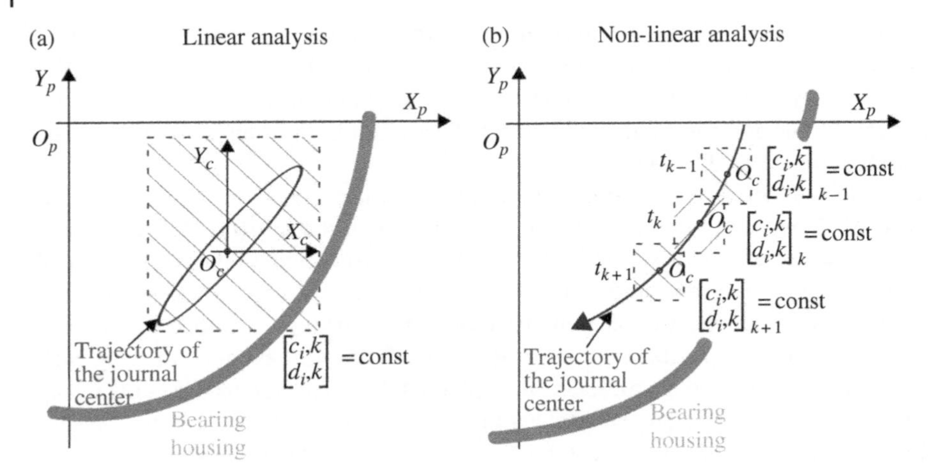

Figure 1.2 Methods of analysis for (a) small and (b) large displacement of the journal.

1.2 Review of the Literature on Numerical Determination of Dynamic Coefficients of Bearings

Attempts at a theoretical description of the operation of rotating machinery have been undertaken since objects with rotating elements began to be used in practice (Chmielniak 1997; Miller 1985; Zieliński 1969). The intensive development of these machines, and with it a more detailed description of them, took place at the turn of the eighteenth and nineteenth century. Significant progress in the theoretical description of rotating machinery began with pioneering works on turbines used for energy conversion (Bannik and Słuczajew 1956; Chodkiewicz 1998; Kiciński and Żywica 2014b; Łączkowski 1974; Nikiel 1956; Smolaga 1959; Wiśniewski 1974). The basis for the theoretical model of hydrodynamic plain bearings is the fundamental equation of hydrodynamic lubrication theory, i.e. Reynolds equation (Reynolds 1886). To this day we encounter some problems, the most important of which are (Barwell 1984): analysis of three-dimensional flow in the bearing lubrication gap, taking into account changes in viscosity, and determining the limits of integration of the pressure distribution curve (which is related to the occurrence of cavitation phenomenon). The impact of the supporting structure (Żywica 2008) is not insignificant either.

The first researcher who suggested applying Reynolds' theory to the calculation of slide bearing dynamics was S. Dunkerley in 1894 (Dunkerley 1894). In 1904 Sommerfeld solved the case of two-dimensional flow in plain bearings on the assumption that the liquid is able to transfer an unlimitedly high pressure (Sommerfeld 1904). Such a simplification caused the obtained result to be incorrect, but a dimensionless parameter determining the hydrodynamic similarity of bearings, the so-called Sommerfeld number, was described. D.G. Christopherson was the first to solve the Reynolds equation using the relaxation method (Christopherson 1941). In his work he stated that assuming that viscosity of oil is constant, it is possible to correctly determine the value of the bearing's load, the amount of oil flowing through the bearing, and temperature changes. The assumption made it impossible to determine the exact friction resistance and load direction. This method (with the application of finite differences) was implemented by C. Cryer in 1971 (Cryer 1971).

H.W. Swift (1932) and W. Stieber (1933) proposed a method of modeling cavitation by applying boundary conditions, currently called Reynolds boundary conditions in the literature. The study of hydrodynamic instability was carried out by, among others, A.C. Hagg and P.C. Warner, who proved that the cause of this phenomenon is oil whirls (Hagg and Warner 1953). Even the most complex bearing models enabled the analysis of the rotor–bearing system only in the range of small vibrations. It is a linearization of systems, which in reality often showed strong non-linear properties. A typical approach is to assume constant values for the stiffness and damping coefficients of the lubricating film, which are determined only for the kinetostatic equilibrium point.

A more detailed examination of the properties of hydrodynamic bearings was made possible with the use of numerical calculation methods. The most commonly used methods are the finite volume method and the finite difference method (Lund 1987; Someya 1989; Wasilczuk 2012; Zienkiewicz and Taylor 2000). The finite element method is most frequently used to analyze the bearing journal displacement (Andreau et al. 2007).

While discussing the development of contemporary computational models, one should mention the considerable achievements of the Turbine Dynamics and Diagnostics Department (formerly the Department of Dynamics of Rotors and Slide Bearings), which is a part of the organizational structure of the Institute of Fluid Flow Machinery of the Polish Academy of Sciences in Gdansk. Under the management of Professor Jan Kiciński, new computational models of plain bearings have been developed and experimental research has been carried out for many years (Kiciński 1994, 1988; Kiciński et al. 1997). It is worth mentioning achievements such as the formulation of a non-linear elastodiatermic model of a plain bearing (Kiciński 1989) and participation in the development of the DT-200 diagnostic system. It was installed on one of the units of the power plant in Kozienice. Also a software package called MESWIR was created. It enables an extensive analysis of bearings and complex rotor lines, even after exceeding the stability limit. The MESWIR package includes, among others, the following programs: NLDW, KINWIR, LDW, TRADYN, DIADEF, ADIAB, ISOSLEW, and DYNWIR. The first three programs were also used during the numerical analysis presented in this monograph. The description of the MESWIR system can be found in Kiciński (2005).

In recent years, numerous attempts have been made to calculate stiffness and damping coefficients in various bearing configurations, for various lubricants (Zhang et al. 2015) and geometries (Illner et al. 2015). New types of bearings and their better numerical description are being developed. The impact of boundary cavitation conditions on dynamic systems with hydrodynamic bearings is described in Daniel et al. (2016). Characteristics describing the operation of a floating ring bearing are presented in Chasalevris (2016).

1.3 Review of the Literature on Experimental Determination of Dynamic Coefficients of Bearings

Due to difficulties in numerical calculation of stiffness and damping coefficients of bearings, many experimental methods have been proposed to determine these coefficients (Dimond et al. 2009; Tiwari et al. 2004). The calculation of stiffness and damping coefficients can be performed in the time or frequency domain. Zhang et al. (1992a, 1992b) as well as Chan and White (1991) determined the stiffness and damping coefficients of two symmetrical bearings by adjusting the curves in the frequency response. This approach assumes a rotor with two identical bearings and allows for the calculation of eight dynamic factors (four stiffness coefficients and four damping coefficients). Since many rotor–bearing systems are not symmetrical and their vibration trajectories have more complex shapes than those for symmetrical systems, the limitation associated with the symmetry of the system in many cases cannot be applied. The method of curve matching also consumes a lot of computational power.

The work of Qiu and Tieu (1997) presents a method extending the calculation of eight coefficients of one bearing, allowing the calculation of a total of 16 stiffness and damping coefficients, for two different hydrodynamic bearings (four stiffness coefficients and four damping coefficients for each bearing). A method of determining the linearized form of coefficients (each stiffness and damping coefficient described by means of a single value) was developed. The authors believe that this method is the most effective method for determining stiffness and damping coefficients. In order to experimentally determine the stiffness and damping coefficients of bearings, it is necessary to force the rotor vibrations. This can be achieved in several ways. The three most commonly used are: impulse excitation of the rotating rotor by means of a modal hammer, the use of additional unbalance of the rotor, and the use of vibration inductors. Qiu and Tieu state in their work that the impulse excitation method is the most economical and convenient way to determine these coefficients. Dynamic coefficients of bearings are calculated on the basis of the response signal of the system measured after inducing the rotor in its central part by means of a modal hammer. The signals in the frequency domain are then used for further calculations. The authors suggest that a wider range of identification should be covered in the calculations in order to ensure greater repeatability of the results.

Tiwari and Chakravarthy (2009) described and demonstrated two different identification algorithms for the simultaneous estimation of the residual unbalance and the dynamic parameters of bearings of a rigid rotor–bearing system. The first method uses the impulse response measurements of the journal from bearing housings in the horizontal and vertical directions, for two independent impulses on the rotor in these directions. Time-domain signals of impulse forces and displacement responses are transformed into the frequency domain and are used for the estimation of the residual unbalance and bearing dynamic parameters. The second method employs the unbalance responses from three different unbalance configurations for the estimation of these parameters. The simulated responses were in fairly good agreement with experimental responses in terms of mimicking predominant resonances. The identified unbalance masses matched quite well with the residual masses taken in the dynamically balanced rotor–bearing test rig.

Meruane and Pascual (2008) described a method of numerical determination of non-linear stiffness and damping factors of hydrodynamic bearings. The calculation of fluid film coefficients using a non-linear method was performed for large displacements in the bearing (20–60% of the bearing gap). The non-linear effect was defined by extending the third-order Taylor equations. The non-linear model was created on the basis of a laboratory test stand. It was found that non-linear properties were revealed with oil vortexes that are not taken into account in classical linear models.

Work on modification of experimental methods in order to obtain greater accuracy is currently in progress. Miller and Howard (2009) described a method for identifying stiffness and damping coefficients by using the extended Kalman filter. This filter was developed to estimate the linearized form of stiffness and damping coefficients of bearings in rotor–bearing systems, taking into account noise and unbalance. The system uses impulse excitation. In this method, bearings are modeled as stochastic, random values using the Gauss–Markov model. The noise part is introduced into the system as an estimation error, including modeling and uncertainty of measurement. The system contains two user-defined parameters that can be used to fine tune the operation of the filter. They refer to the covariance of the system and the noise variables. The filter was tested using numerically created data of a system of two identical bearings, reducing the number of unknown coefficients to eight. The method was used to determine the main dynamic coefficients of bearings, while the cross-coupled coefficients were determined with a lower accuracy.

Particular difficulties in numerical determination of dynamic coefficients of bearings may be caused by bearings of complicated construction, e.g. foil bearings (Kiciński and Żywica 2012). The results of experimental identification of dynamic coefficients of a large foil bearing are presented in Wang and Kim (2013). Dynamic characteristics of a hybrid foil bearing (hydrodynamic + hydrostatic), 101.6 mm in diameter and 82.6 mm in length, are presented in that work. The stiffness coefficients were determined using two methods: the quasi-static method by determining deflection curves in the time domain and the impulse method in the frequency domain. The values of damping coefficients were determined using the impulse method only. The values of stiffness coefficients determined using the two above methods were similar, with the differences of around 4–7 MN/m depending on the speed, load, and supply pressure. Frequency calculations were characterized by greater discrepancies in the obtained results. As part of the work, a numerical model was also created, using the linear perturbation method. On the basis of the shaft deflection it was found that the results were similar to those of the experimental research.

A paper by Delgado (2015) presents calculations for a hybrid gas bearing (Kazimierski and Krysiński 1981) of a complex construction, characterized by rigid geometry and complex foil construction. The bearing operates on two lubrication films: hydrostatic and hydrodynamic. Such a procedure ensures the generation of appropriate load-bearing capacity and stiffness of the entire system. The paper presents an experimental verification of stiffness and damping coefficients for a bearing with a diameter of 110 mm. The results were obtained on a specially designed laboratory test rig. The variable parameters during the experiment were: hydrostatic supply pressure, driving force frequency, and rotor speed. Experimental research was aimed at evaluating the application of this bearing type in large-scale energy conversion machines. The dynamic tests showed poor sensitivity of the main stiffness coefficients to most of the test parameters. The frequencies and speeds were an exception: the higher the speed and frequency of the driving force, the lower the value of the calculated stiffness coefficients.

A paper by Kozánek et al. (2009) presents the results of experimental calculations of aerostatic radial bearings on the Bentley Nevada laboratory test rig. Various types of bearings as well as their static and dynamic characteristics were examined. Different methods of identification of dynamic coefficients of bearings were applied. Only the main stiffness and damping coefficients were calculated. The impact of cross-coupled stiffness and damping parameters was examined on the basis of numerical simulations.

During calculations, the mass matrix was defined as a matrix of known parameters, stiffness and damping matrices were determined. The authors recommended that when conducting experimental research the main values of stiffness and damping coefficients should be determined first, followed by cross-coupled ones, which are more susceptible to errors. According to the authors, increasing the bearing supply pressure and frequency of driving force negatively affects the correctness of the obtained values of dynamic coefficients of bearings.

Another example of experimental determination of parameters of aerostatic bearings is presented in Kozánek and Půst (2011). As part of the paper, a numerical model was developed to calculate the stiffness and damping parameters of both radial and thrust bearings.

A new method of identification of stiffness and damping of a bearing, based on phase plane diagrams, is presented in a paper by Jáuregui et al. (2012). The authors emphasize that reliable determination of dynamic coefficients of bearings is a huge challenge, particularly in non-linear systems. They also claim that a single, universal mathematical model does not exist, and the identification of parameters of a system depends on the measured data and the reference model. This model of phase plane diagrams works well when the coefficients are strongly dependent on frequency.

Experimental calculations of stiffness and damping parameters of bearings are not only done for radial bearings, but also for thrust bearings. The experimental determination of the parameters of stiffness and axial damping of foil bearings are presented in a paper by Arora et al. (2011). The paper presents a diagram of the procedure of identification of these coefficients, a description of the laboratory test rig, and a diagram of the values of stiffness coefficients in the function of rotational speed. The Monte Carlo algorithm was used to determine the dynamic parameters of the rotor supported on the magnetorheological layer of the lubricating film (Zapomel et al. 2014).

The method of impulse excitation for the determination of bearing coefficients, which is used in this monograph, is carried out in the frequency domain. Its first basic version was proposed by Nordmann and Schoelhorn (1980). Qiu and Tieu extended the calculation algorithm with the possibility of calculating 16 stiffness and damping coefficients (Qiu and Tieu 1997). This monograph describes a modification of the algorithm developed by Qiu and Tieu to calculate the damping coefficients and stiffness of bearings (using the impulse method and approximation with the least squares method). The algorithm has been extended by the possibility of calculating eight mass coefficients. Determination of damping, stiffness and mass coefficients using a single algorithm enables verification of the results at the initial stage of operation. Since the mass of the shaft is usually a known size, the correctness of the determined dynamic coefficients of bearings can be verified on the basis of mass coefficients. This approach makes it possible to determine all dynamic parameters of the rotor–bearing system through experimental research.

1.4 Purpose and Scope of the Work

The analysis of the literature and research conducted earlier in the Institute of Fluid Flow Machinery of the Polish Academy of Sciences confirm that dynamic coefficients of bearings have a key influence on the dynamic properties of rotating machinery. Since experimental methods for determining the stiffness and damping coefficients of bearings are usually described in the literature as burdened with large calculation errors, and the range of applicability of non-linear numerical methods (Bonet and Wood 2009; Sathyamoorthy 2000) is not clearly defined, the following objectives of work were formulated. They correspond with the subject of my dissertation (Breńkacz 2016) and over a dozen scientific articles.

The main aim of the work is to develop and describe a method of experimental determination of dynamic coefficients of hydrodynamic plain bearings and its verification. An additional objective of the work is to determine dynamic coefficients of bearings on the basis of experimental and numerical studies (linear and non-linear) in a wide range of rotational speeds of the rotor, taking into account resonance and speeds higher than the resonant speeds. Achievement of these two objectives will make it possible to compile data necessary to determine the ranges of linear and non-linear adequacy of methods for determining the dynamic coefficients of hydrodynamic bearings on the example of the rotating machinery under study.

The author participated in a one-year internship (from April 2013), which took place at LMS International in Belgium. The internship was organized as part of the STA-DY-WI-CO (European Commission/CORDIS 2019) project, which is part of the Marie Curie IAPP (Industry Academia Partnerships and Pathways) program. LMS International (www.plm.automation.siemens.com) is a leading manufacturer of instrumentation and software for measuring sound and vibration. This company cooperates with Samtech, the producer of Samcef Rotors software for analyzing the dynamics of rotors (https://blogs.sw.siemens.com/simcenter/). The issues described in this monograph were a common subject of the work of the two aforementioned companies.

In order for the basic objectives of the work to be achieved, it is necessary to carry out a number of intermediate tasks. The most important include:

- Development of a calculation algorithm which will make it possible to experimentally determine the stiffness, damping and mass coefficients of two hydrodynamic radial bearings.
- Verification of the developed algorithm.
- Evaluation of the sensitivity of the experimental method for the determination of stiffness, damping and mass coefficients.
- Conducting experimental tests in order to determine the basic dynamic characteristics of the tested laboratory test rig.
- Conducting experimental studies on the basis of which dynamic coefficients of two hydrodynamic radial bearings will be determined.
- Verification of the calculated experimental stiffness and damping coefficients based on the numerical model using the Abaqus software.
- Development of numerical models of the rotor and plain bearings with the use of the MESWIR software series. These programs enable the calculation of dynamic coefficients of hydrodynamic bearings using linear and non-linear algorithms.
- Numerical calculation of dynamic coefficients of bearings using linear and non-linear numerical models.
- Verification of the values of dynamic coefficients of bearings obtained with the use of numerical methods, by comparing the changes in journal vibration amplitudes and journal vibration trajectories (calculated on the basis of numerical analyses) with the results of experimental tests.

- Comparison of dynamic coefficients of bearings calculated using three different methods (on the basis of linear algorithm and experimental tests as well as linear and non-linear numerical models) and determination of the ranges of adequacy of their application.

All the above objectives were achieved and are described in this book. This chapter provides an introduction to the dynamics of rotating machinery. It presents the subject matter tackled in this monograph on a background of research carried out worldwide. Fundamental issues concerning dynamic properties of rotor–bearing–supporting structure systems are also discussed and a review of the literature on numerical and experimental methods of determining dynamic properties of rotating machines is presented. The basic research problem is formulated, and the aim of the study is defined.

Chapter 2 shows practical applications of bearing dynamic coefficients. Based on a simple example, it is shown how stiffness and damping coefficients affect dynamical systems. Other examples demonstrate how static and dynamic forces act. Basic equations for the vibrating motion are presented. The following examples show a dynamical system with one and two degrees of freedom. A practical example shows what influence cross-coupling coefficients have and what they mean.

Chapter 3 contains a description of the laboratory test rig and the characteristics drawn up based on experimental research. Rotor vibrations during run-up, acceleration of vibrations of the bearing support and modal analysis enabled the preparation of a complete picture of the dynamics of rotor machines. Vibrations and vibration trajectories of the journal were used to verify the numerical analyses carried out.

Chapter 4 discusses the research tools used during the implementation of the work. Accelerometers were used to measure vibrations of the structure, eddy current sensors were used to measure rotor displacement, and a laser tachometer was used to measure rotational speed. Measurement data were archived with Scadas Mobile and processed with LMS Test. Lab software. The analysis of rotor dynamics was carried out using a Samcef Rotors program and the MESWIR software series. The algorithm for calculating dynamic coefficients of bearings on the basis of experimental tests was developed using the Matlab program. In the same program, operations were performed on signals from numerical and experimental research. Verification of experimental results was carried out using Abaqus software.

Chapter 5 presents the method of experimental determination of dynamic properties of rotating machines. Under one algorithm, mass coefficients were also determined along with stiffness and damping coefficients. The mass coefficients can be interpreted as the mass of the shaft involved in vibrations (Kruszewski et al. 1996; Kruszewski and Wittbrodt 1992). Comparing the known mass of the shaft with the calculated mass coefficients, conclusions can be drawn on the correctness of the calculated stiffness and damping coefficients.

Chapter 6 presents one of the most time-consuming stages of the process related to the experimental determination of stiffness, damping and mass coefficients, i.e. appropriate signal preparation. It is necessary to select only a part of the signal, find a reference signal showing stable operation and subtract these signals. Dynamic coefficients of bearings were calculated on the basis of an appropriately prepared signal.

The sensitivity analysis of the method of experimental determination of dynamic coefficients of bearings is presented in Chapter 7. Using a numerical model, the influence of six parameters on the results of the calculations is presented. The analysis of the impact of these parameters is carried out on the basis that experimental research would not be possible or would be burdened with a significant error.

Chapter 8 presents the experimental tests, on the basis of which the stiffness, damping and mass coefficients were determined. The measured journal vibration signals, driving force signals and calculation results are presented.

Chapter 9 describes the method of linear and non-linear numerical calculations and numerical calculation models. Dynamic coefficients of hydrodynamic bearings were determined on their basis. The results of numerical calculations and their verification are presented.

The penultimate chapter (Chapter 10) summarizes the calculated stiffness and damping coefficients obtained using three different calculation methods. The differences between the results of linear experimental calculations and linear and non-linear numerical calculations are presented. Comments are made on the reasons for the discrepancies obtained.

The final chapter (Chapter 11) contains a summary of the results obtained during the research and conclusions from the work carried out.

2

Practical Applications of Bearing Dynamic Coefficients

Regardless of the type of bearings, be it rolling-element, gas, magnetic, or hydrodynamic, we assume that the dynamic coefficients of bearings determine the rotordynamics (in addition to other factors such as the rotor design, the material it is made of, the properties of the supporting structure, etc.). Their values result directly from the properties of the bearing. Due to the nature of operation of different types of bearings, they are characterized by stiffness and damping coefficients falling within different ranges. The conditions in an oil-based hydrodynamic bearing and in a rolling-element bearing are sufficiently different to result in different dynamic coefficient values. Higher damping value can usually be found in hydrodynamic bearings, while higher stiffness coefficient values are typical in rolling-element bearings. At this point it should also be stressed that the dynamic coefficients of bearings should be treated as an "abstract creation" of sorts, which is often impossible or very difficult to measure directly. It is also often connected with large error. With correctly determined values of stiffness and damping coefficients of bearings, the created numerical models reflect the results of experimental research very accurately. The creation of such models enables the design of a properly functioning rotor–bearing–supporting structure system.

Most commercial applications use a linear form of stiffness and damping coefficients, i.e. they do not change over time and are not described by a function dependent on the excitation force. They may have fixed values for certain rotational speeds, but they do not change when the rotor is operating at constant speeds.

The rotor used to analyze the influence of dynamic coefficients of bearings is shown in Figure 2.1. It is a symmetrical rotor with a disk in the center. Its diameter is Φd and its length is L. The second-order differential equation (2.1) is most commonly used to describe it. The finite element method is the most commonly used calculation method. It can be used to develop stiffness (K), mass (M), and damping (C) matrices reflecting the geometry and properties of the system and an excitation force matrix (F) containing the excitations affecting the system. Depending on the type of finite elements (beam elements, three-dimensional elements, etc.), a numerical model is created, whose individual matrix elements are arranged in an appropriate manner.

$$M\ddot{x}(t) + (C + \Omega G)\dot{x}(t) + Kx(t) = F(t) \tag{2.1}$$

Bearing Dynamic Coefficients in Rotordynamics: Computation Methods and Practical Applications,
First Edition. Łukasz Breńkacz.

Companion website: www.wiley.com/go/brenkacz/bearingdynamiccoefficients

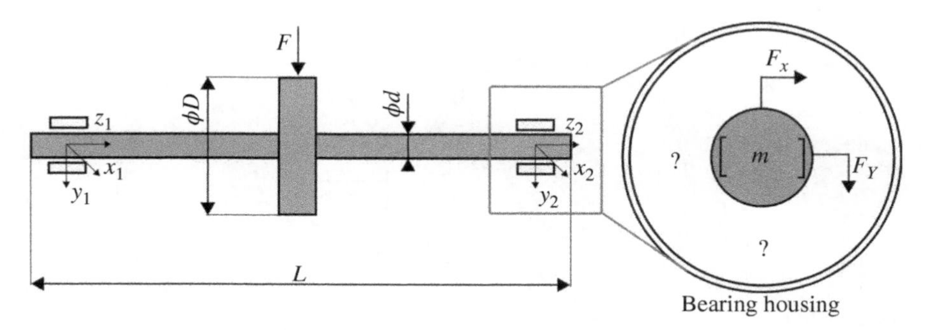

Figure 2.1 Bearing as part of a rotating system. Question marks indicate the places where dynamic coefficients of bearings occur.

where M is the mass matrix, C is the damping matrix, ΩG is the speed-dependent gyroscopic torque matrix, K is the stiffness matrix, and $F(t)$ is the excitation force.

The fundamental question in the analysis of the bearing concerns the manner of considering the relationship between the journal and the bearing housing. The contact area is illustrated with question marks in Figure 2.1. When considering the differential equation (2.1) of a singular point, the radial bearing, the stiffness and damping coefficients can be defined by the equations formulated in Chapter 1, namely Eqs. (1.2), (1.3), (1.4), and (1.5).

A simplified model was used to analyze the influence of stiffness and damping coefficients on the rotordynamics. It does not provide the opportunity for in-depth analysis as the bearing is a complex subject of research which has been carried out for decades in a number of research centers. The presented model, on the other hand, provides a very good overview of the phenomena that largely determine the operation of bearings and rotors. The motion of the bearing journal is treated as a point motion of concentrated mass, which is equal to half of the rotor's mass. This is a very good approximation of bearing operation assuming that the symmetrical rotor is supported on two radial bearings and during its operation no influence of the coupling or any other interference is observed. For the purpose of this analysis it was assumed that the gyroscopic effect occurring in the system is negligible ($G = 0$) and that there is no gravitation. With such assumptions, Eq. (2.1) can be treated as an equation of bearing movement in which values K and C will be the recorded stiffness and damping coefficients of bearings k and c, respectively. The M matrix corresponds to the mass m of the section of the shaft supported by the radial bearing.

Differential equations for the successive analyzed systems from Figure 2.2 are presented in the following. Equation (2.2) describes the movement of mass from Figure 2.2a, Eq. (2.3) describes the movement of mass from Figure 2.2b, and Eq. (2.4) describes the movement of mass from Figure 2.2c. All these cases are analyzed in the following three subsections. Such division will allow the influence of particular elements of bearing design to be presented.

$$m_{yy}\ddot{x}(t)+c_{yy}\dot{x}(t)+k_{yy}x(t)=F_y(t) \tag{2.2}$$

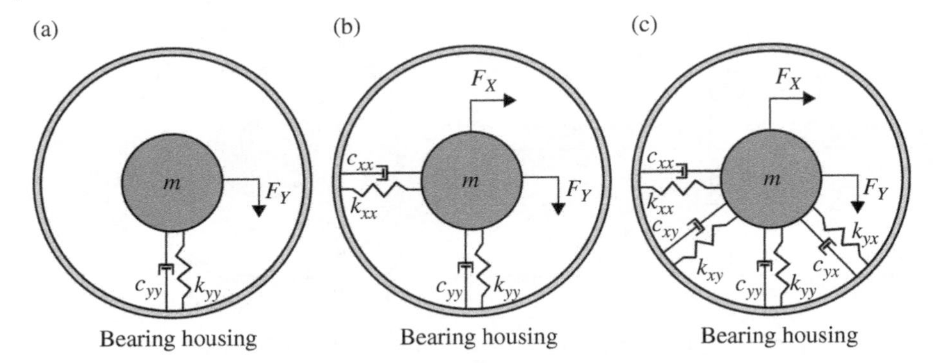

Figure 2.2 Different bearing models: (a) oscillating mass with one degree of freedom; (b) bi-directional oscillating mass (two degrees of freedom); and (c) bi-directional oscillating mass with the inclusion of cross-coupled stiffness and damping coefficients.

$$\begin{bmatrix} m_{xx} & \\ & m_{yy} \end{bmatrix}\ddot{x}(t)+\begin{bmatrix} c_{xx} & \\ & c_{yy} \end{bmatrix}\ddot{x}(t)+\begin{bmatrix} k_{xx} & \\ & k_{yy} \end{bmatrix}x(t)=\begin{bmatrix} F_x(t) \\ F_y(t) \end{bmatrix} \tag{2.3}$$

$$\begin{bmatrix} m_{xx} & m_{xy} \\ m_{yx} & m_{yy} \end{bmatrix}\ddot{x}(t)+\begin{bmatrix} c_{xx} & c_{xy} \\ c_{yx} & c_{yy} \end{bmatrix}\dot{x}(t)+\begin{bmatrix} k_{xx} & k_{xy} \\ k_{yx} & k_{yy} \end{bmatrix}x(t)=\begin{bmatrix} F_x(t) \\ F_y(t) \end{bmatrix} \tag{2.4}$$

In the next subsection, the results of the analysis are shown for constant value and sinusoidal excitation force. Comparison of sample results allows differences in displacement generated by different types of forces to be indicated.

Sinusoidal excitation was analyzed in two variants. In the first case the value of force amplitude was constant for the whole range of analyzed frequencies – it was multiplied by the sinusoidal function. In the second case, the amplitude of the force generated by the unbalance increased along with the rotational speed (this was also multiplied by the sinusoidal function). The latter case of calculation presents the real load on the rotor, but due to various simplifications in the literature on oscillation analysis, most commonly results are provided only for one variant.

2.1 Single Degree of Freedom System Oscillations

The analysis of oscillations can be started from a single degree of freedom system. There are countless books and studies devoted to this issue. The purpose of this chapter is not to quote an appropriate theory, but only to provide the necessary introduction for further analysis. Normally, consideration of a single degree of freedom system (mass m attached to a fixed surface) starts with a theoretical system representing the free oscillation of a spring (without a damper), the movement of which is described by Eq. (2.5).

$$m\ddot{x}(t)+kx(t)=F(t) \tag{2.5}$$

The solution of this equation presented using the formula shown in (2.6) indicates that after the initial (e.g. impulse) excitation, the system will perform an oscillating motion with a natural frequency marked as ω_0. The angular frequency is usually expressed in radians per second, as is the case with the formula shown. The unit can be converted to hertz (Hz) by multiplying the formula shown in (2.6) by $1/2\pi$. Each oscillation system has one or more natural frequencies. This simple relationship can be used for a general understanding of what happens to more complex systems when mass or stiffness is included. The formula (2.6) explains, for example, why, when a car or truck is fully loaded, its suspension works differently (e.g. it is more sensitive) when compared with driving the same vehicle without a load. This is due to the fact that after an increase in mass, the natural frequency of the system's own oscillations decreases.

$$\omega_0 = \sqrt{\frac{k}{m}} \tag{2.6}$$

A damper can be added to the previously presented system which did not feature any damping. The differential equation of motion will change from (2.5) to (2.2). Double indexes have been used for consistent results of subsequent recordings of stiffness and damping coefficients. Imagine adding damping as adding a force proportional to the speed of the mass. Damping is called viscous because it models the action of a liquid inside an object. The constant proportionality c is called the damping coefficient and uses the units of force divided by speed (N·s/m).

The oscillation frequency of the damping system can be recorded according to formula (2.7). It is lower than the natural frequency of the oscillator without damping.

$$\omega = \sqrt{\omega_0^{\,2} - \left(\frac{c}{2m}\right)^2} \tag{2.7}$$

From the point of view of system dynamics analysis, a very important value is the critical damping c_c defined according to formula (2.2):

$$c_c = 2\sqrt{km} \tag{2.8}$$

The solution of the differential equation (2.2) depends on the amount of damping. To describe the behavior of a damped oscillation system, a damping factor is introduced, ζ (zeta). The value of damping ζ determines the behavior of the system. Damped harmonic oscillator systems can be divided into four categories depending on the value of the ζ parameter defined by Eq. (2.9):

- **Heavy damping** ($\zeta > 1$) – the system does not oscillate but follows the exponential decay till it reaches equilibrium. The higher the value of damping ζ, the slower the changes in the system.
- **Critical damping** ($\zeta = 1$) – the system returns to equilibrium without oscillation, and this is the quickest way to achieve equilibrium without oscillation. The damping for this case is defined by formula (2.8).
- **Low damping** ($0 < \zeta < 1$) – the system oscillates with exponentially decreasing amplitude and frequency lower than that of an undamped system. The increase in damping

results in a faster amplitude decay and a decrease in the frequency of the system's own oscillations.

- **Undamped** ($\zeta = 0$) – the system performs oscillations of a constant amplitude at its natural resonance frequency.

$$\zeta = \frac{c}{2\sqrt{km}} \tag{2.9}$$

The damping factor ζ is a dimensionless parameter. For example, in metal structures, such as aircraft fuselages or engine crankshafts, its value may be less than 0.05, while values for car suspensions are in the range 0.2–0.3.

2.1.1 Constant excitation Force

The first analyzed system with one degree of freedom (Figure 2.2 and described by Eq. (2.2)) can be loaded with static and dynamic (variable over time) force. The static force load will be presented in this subsection. Of course, this is not an excitation characteristic of rotordynamics calculations, but it makes it possible to illustrate the nature of the system's response in the simplest type of excitation. The influence of dynamic force will be presented in the next subsection.

Dynamic coefficients of bearings are sometimes called "bearing force coefficients." The name rightly suggests that they are dependent on the excitation force acting on the system. If a constant force is applied, the system will change as shown in Figure 2.3.

In order to visualize the sample results, a system with one degree of freedom was built. The parameters adopted for mass, damping, and stiffness are:

- $m = 1.5\,\text{kg}$
- $c_{yy} = 6421\ \text{N·s/m}$
- $k_{yy} = 6.63 \cdot 10^6\ \text{N/m}$

Figure 2.4 presents the results for a constant force load of 6 N. The entire force value was applied in the first time interval. The displacement of the mass in the direction of the force acting is observed (black curve with square-shaped symbols). The next two curves were generated by including two and four times larger damping coefficients in the system. The

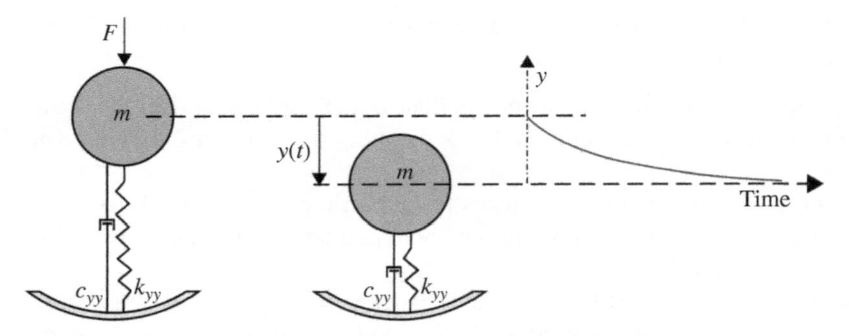

Figure 2.3 Displacement of the mass attached to a fixed support by means of stiffness and damping coefficients as a result of a constant force.

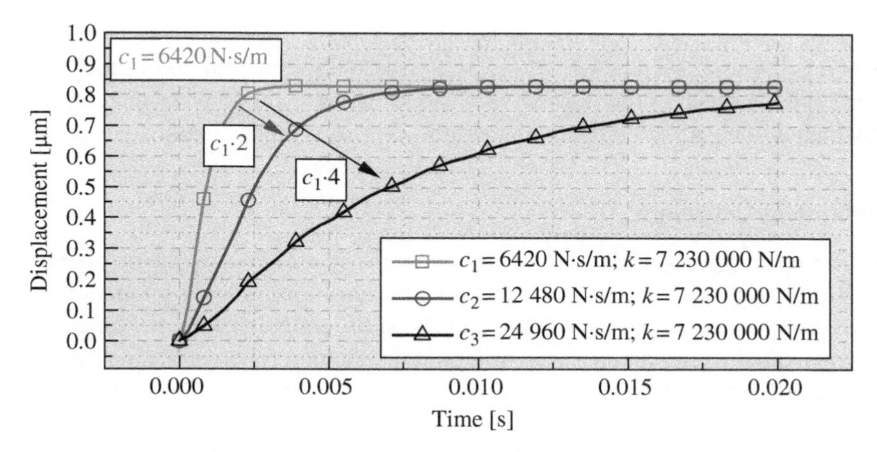

Figure 2.4 The effect of a change in the damping coefficients on the displacement of the mass point which is being acted on by a constant force. The arrows represent the direction of displacement change due to the increase of damping coefficient values.

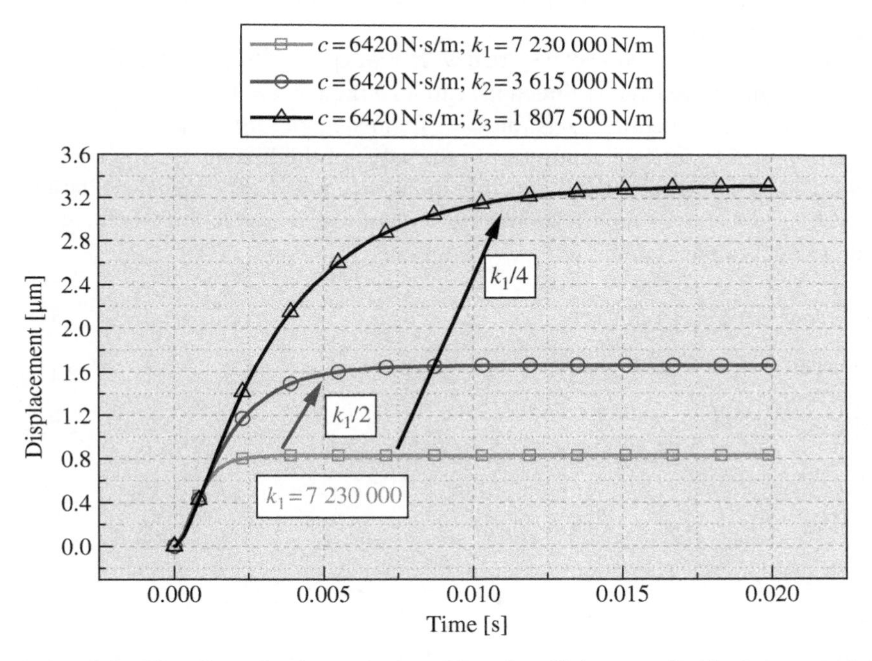

Figure 2.5 The effect of a change in the stiffness coefficients on the displacement of the mass point which is being acted on by a constant force. The arrows represent the direction of displacement change due to the decrease of stiffness coefficient values.

chart shows that with the increase of the value of the damping coefficients, the time during which the mass is displaced increases. After a long enough time, in each case the mass will be displaced to the same level.

The influence of changes in stiffness coefficients on the response of a system with one degree of freedom is presented in Figure 2.5. The black curve with square-shaped symbols was generated for the same data as in the previous case. The next two curves were generated by reducing the stiffness coefficients two (to 3615000 N/m) and four times (to 1807000 N/m). It is clear that as the stiffness coefficients increase, the displacement of the mass point, which is being acted on by the same force, decreases proportionally.

2.1.2 Excitation by Unbalance

For rotors operating at supercritical speeds (higher than the speed at which the first bending form of natural oscillation occurs), after creating the amplitude chart as a function of speed, an increase and then a decrease in the amplitude of oscillations can be observed. This phenomenon is called resonance; it can also be observed for a system with one degree of freedom. The frequency at which resonance occurs can be determined from the relationship (2.7).

In rotordynamics, the unbalance is usually forcing, which is modeled by means of an unbalance mass (m_u) rotating at a frequency equal to the rotor speed (ω), which is fixed on a radius equal to r. An illustration of the unbalance model and its operation is presented in Figure 2.6. Considering the movement in one plane, the system's response in the form of a sinusoidal signal is obtained.

The differential equation of motion of a system with one degree of freedom together with the force from the unbalance (in one direction) can take the form of (2.10). If a constant value of excitation force is assumed, regardless of frequency, the force F_u is defined by Eq. (2.11). If the value of the force changes with the value of the frequency squared, it is described by Eq. (2.12). This method of applying unbalance reflects the actual excitation best, however, due to various simplifications in the literature, dynamic force defined by both (2.11) and (2.12) can be found.

$$M\ddot{x}(t) + C\dot{x}(t) + Kx(t) = F_u \tag{2.10}$$

$$F_u = F\cos\omega(t) \tag{2.11}$$

$$F_u = m_u r\omega^2 \cos\omega(t) \tag{2.12}$$

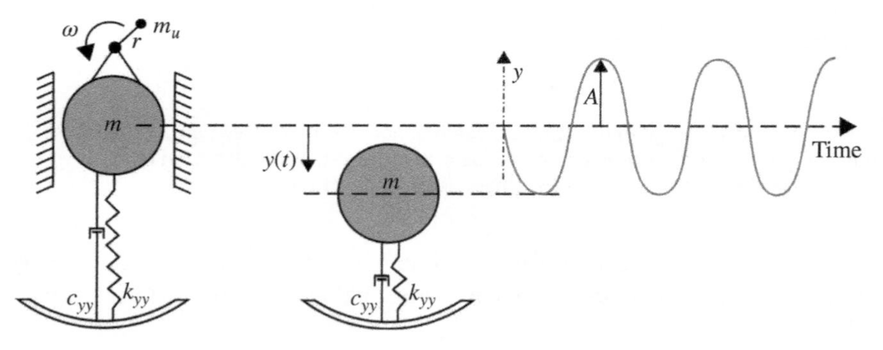

Figure 2.6 A single degree of freedom system with harmonic excitation.

The following data were generated for the same stiffness and damping coefficients: $m = 1.5\,\text{kg}$, $c = 6420\,\text{N/m}$, and $k = 7\,230\,000\,\text{N/m}$. Instead of a constant force value of $F = 6\,\text{N}$, a sinusoidal function with an amplitude of $F = 6\,\text{N}$ and a frequency of $\omega = 300\,\text{Hz}$ (1884 rad/s) was assumed. As a result, a system response presented as a black line with square-shaped symbols in Figure 2.7 was obtained. As in the case shown in Figure 2.4, the next two lines were generated by doubling and quadrupling the damping coefficients. It is clear that the oscillation amplitude decreases with the increase of damping, but these changes are not as directly proportional as in the previous example using constant force.

A very interesting result is obtained by analyzing the changes of stiffness coefficients in the same range as those considered in Figure 2.5 for a force of constant value. The stiffness of the system was reduced two and four times from 7 230 000 N/m, through 3 615 000 to 1 807 500 N/m. After changing the values of stiffness in the analyzed range (Figure 2.8), we observe almost no changes in the system with one degree of freedom with dynamic excitation.

It turns out that the influence of stiffness and damping coefficients is different depending on the excitation force frequency in relation to the resonance frequency. The full picture of the relationship is obtained by creating charts showing the changes in the amplitude of displacement as a function of the increasing frequency of the excitation force. These results can be obtained by using a constant force or a force changing in accordance with the relationship (2.13):

$$F = m_u r \omega^2 \tag{2.13}$$

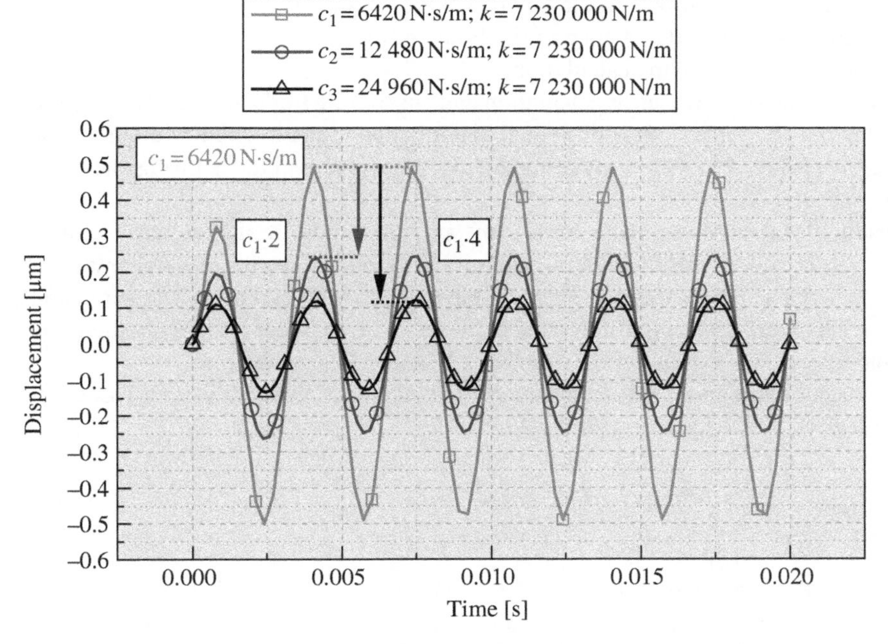

Figure 2.7 The effect of a change in the damping coefficients on the displacement of the mass point which is being acted on by a constant harmonic with a frequency of 300 Hz.

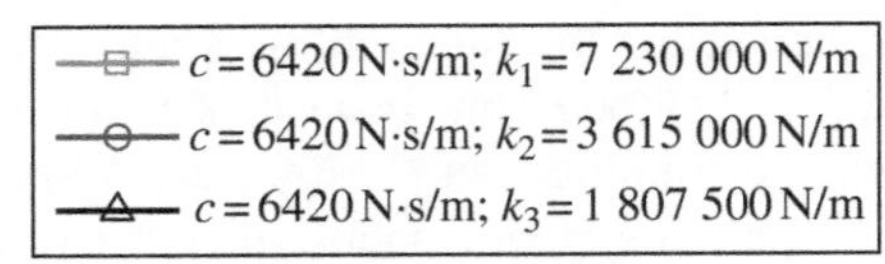

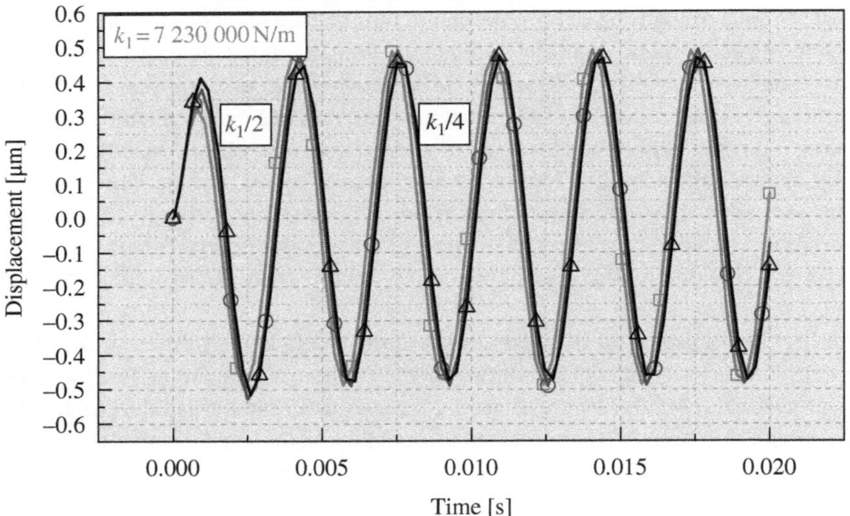

Figure 2.8 The effect of a change in the stiffness coefficients on the displacement of the mass point which is being acted on by a constant harmonic with a frequency of 300 Hz.

where m_u is the unbalance mass, r is the unbalance radius, and ω is the frequency range of the shaft rotation (and thus the change in the frequency of the excitation force as unbalance of the shaft) included in the calculation.

The following values were used as calculation data: $m_u = 0.18 \cdot 10^{-3}$ kg, $r = 17 \cdot 10^{-3}$ m, and $\omega = 0$–4000 rad/s. The force chart is presented as a function of the frequency in Figure 2.9. A constant force value of 13.49 N was assumed. This is the force generated during operation at resonant speed.

As a result of calculations made for a constant amplitude of the excitation force, Figure 2.10 was created. After taking into account the variable force in the whole range of analyzed frequencies, the results presented in Figure 2.11 were obtained. The static and dynamic force had the same amplitude value at the resonant frequency, therefore the value of the resonant oscillation amplitude in Figures 2.10 and 2.11 are the same.

In order to generate Figures 2.10 and 2.11 a single degree of freedom system was used, which is described by the following parameters:

- $m = 1.5$ kg
- $c = 321$ N·s/m
- $k_{yy} = 6.63 \cdot 10^6$ N/m

For each frequency, the amplitude of the stabilized response was recorded and the results (of constant excitation force) are compared in Figure 2.10. The results for a force with variable amplitude which changes along with the frequency according to the formula (2.13) are summarized in Figure 2.11. The value of the force changed as presented in Figure 2.10. The

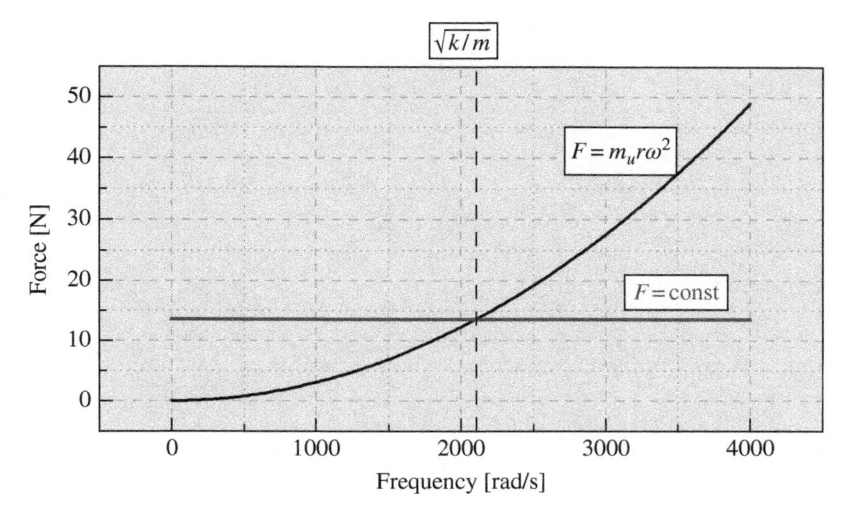

Figure 2.9 Excitation force of constant value and of variable value changing along with frequency.

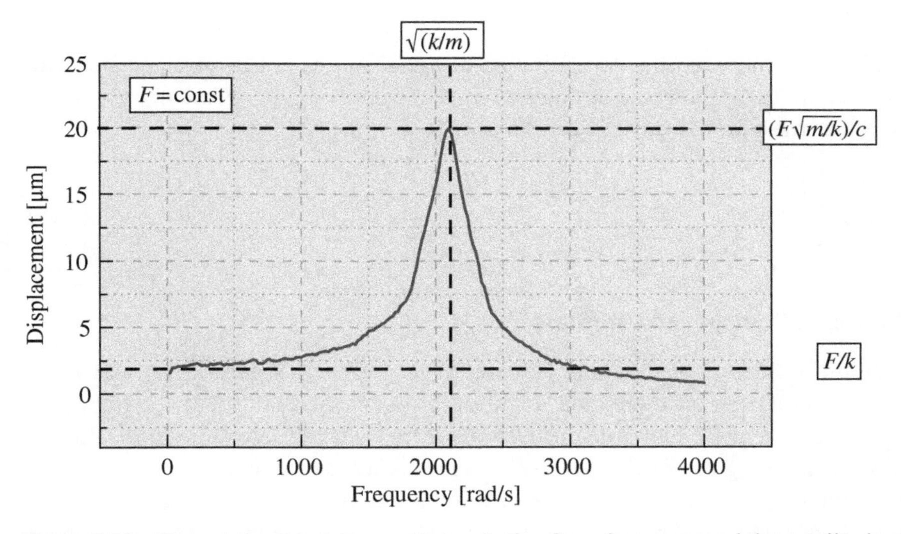

Figure 2.10 The relationship between the excitation force frequency and the amplitude of the system response for a constant force F.

result is a relationship between the frequency of the excitation force and the amplitude of the system response.

The resonance oscillation amplitude for the case with constant force (Figure 2.10) can be determined according to the following relationship (2.14).

$$A_{kr} = \frac{F\sqrt{\frac{m}{k}}}{c} \tag{2.14}$$

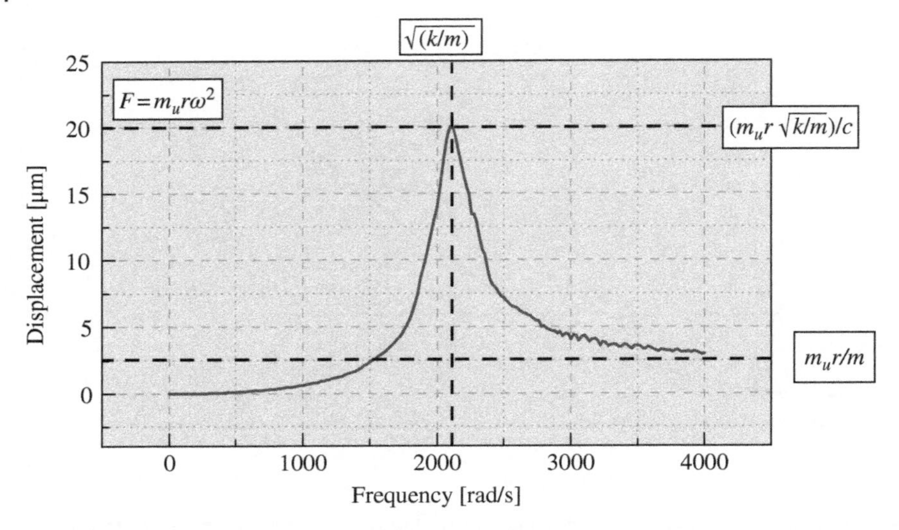

Figure 2.11 The relationship between the excitation force frequency and the amplitude of the system response for the force F variable along with frequency.

The maximum amplitude value for excitation by unbalance can be determined from the formula shown in (2.1). The value of this amplitude depends on the unbalance mass and radius as well as the stiffness, damping and mass coefficients.

$$A_{kr} = \frac{m_u r \sqrt{\frac{k}{m}}}{c} \tag{2.15}$$

2.1.3 Impact of Damping and Stiffness

Analysis of the influence of damping (Figure 2.12), stiffness (Figure 2.13), and mass (Figure 2.14) coefficients on oscillation amplitudes was carried out for the system presented earlier. The analysis took into account the force with variable amplitude and frequency. As the damping increases, a decrease in resonance oscillations was observed. As the stiffness increases, both the frequency of resonant oscillations and the amplitude of these oscillations increased. The opposite effect can be observed when increasing the mass of the system. As the frequency of resonant oscillations decreases, the amplitude of these oscillations is also reduced.

The impact of stiffness, damping and mass coefficients are summarized in Figure 2.15. It shows that in this calculation, increasing the damping coefficients causes a decrease in the amplitude of resonant oscillations. Increased stiffness coefficients result in an increase in the frequency and amplitude of resonant oscillations. Increasing the mass coefficients results in a decrease in the frequency and amplitude of resonant oscillations. Both the mass and stiffness coefficients have a visible impact in the range of higher excitation force frequencies.

A slightly different combination can be found in the literature, showing the influence of stiffness and damping coefficients presented in Figure 2.16. It shows that the damping

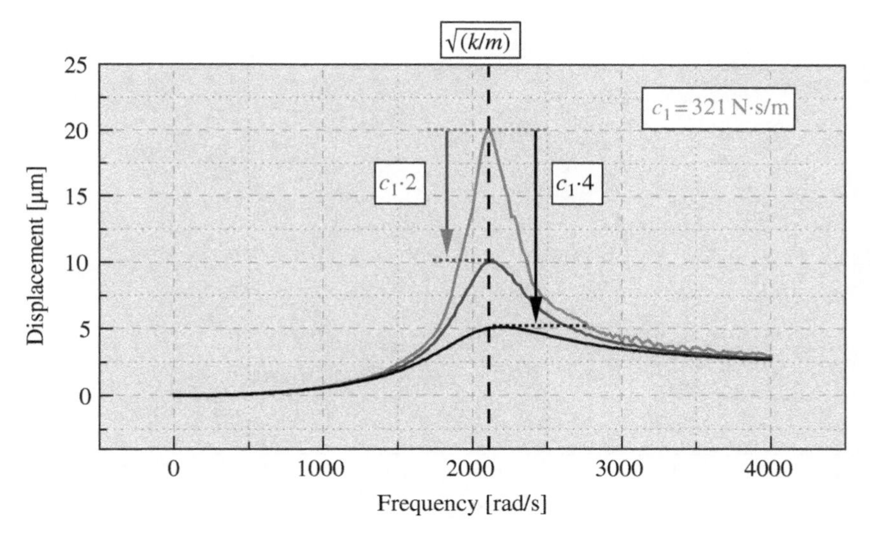

Figure 2.12 Impact of damping coefficients.

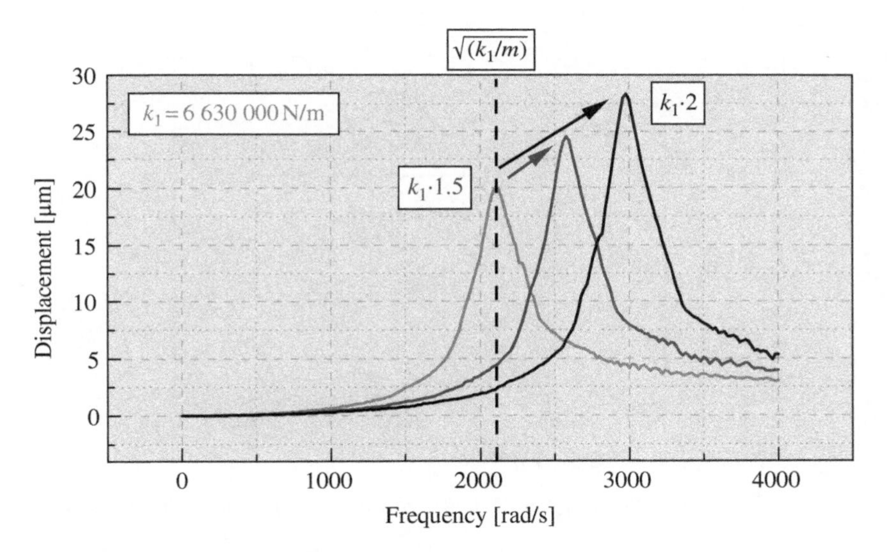

Figure 2.13 Impact of stiffness coefficients.

coefficients influence the results in the resonance frequency range. The stiffness coefficients mainly have impact in the frequency range below the resonant frequencies, while the mass coefficients have impact in the frequency range above the resonant frequencies. This comparison is valid (especially in the range of high and low frequencies compared with the resonance frequency) if a constant value of excitation force is assumed, thus not reflecting the actual excitation of the rotor.

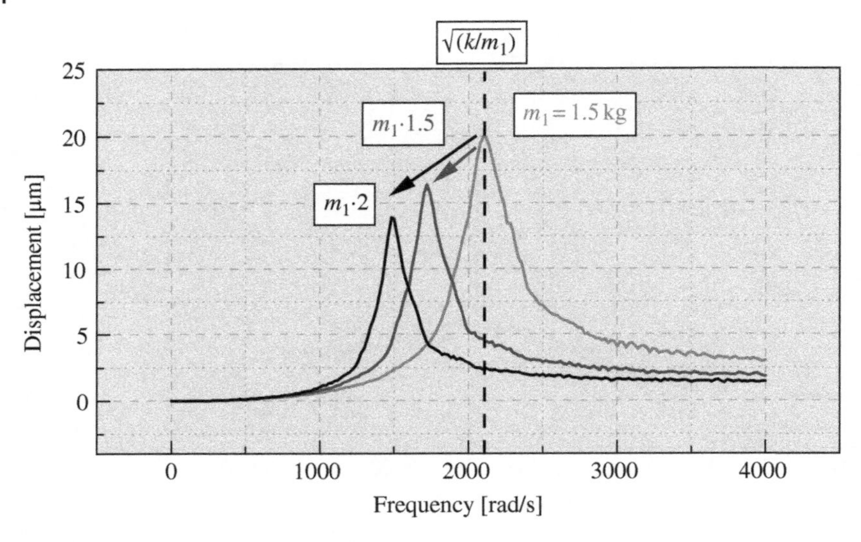

Figure 2.14 Impact of mass coefficients.

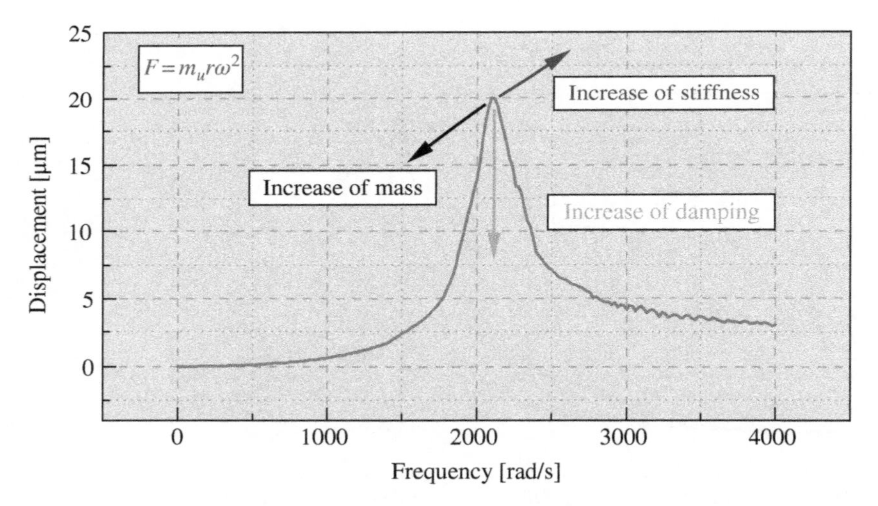

Figure 2.15 Areas of influence of different types of coefficients for variable amplitude values of excitation force.

2.2 Oscillation of Mass with Two Degrees of Freedom

For the analysis of a system with two degrees of freedom, excitation forces acting in two perpendicular directions, of equal frequency and amplitude and shifted in phase are assumed. The source of the excitation force is unbalance. A schematic of a system in which the unbalance m_u rotates counterclockwise at a constant angular velocity ω around radius r is shown in Figure 2.17. The angle between OP and the horizontal axis can be calculated

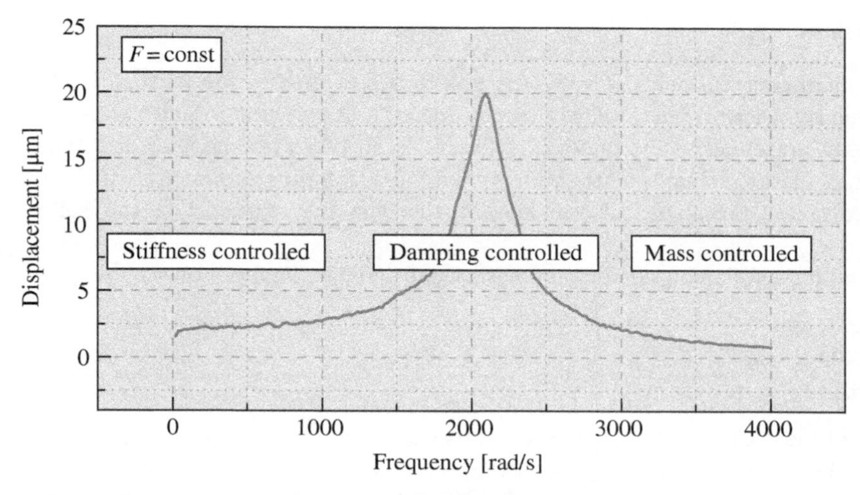

Figure 2.16 Areas of influence of different types of coefficients for the constant amplitude of excitation force.

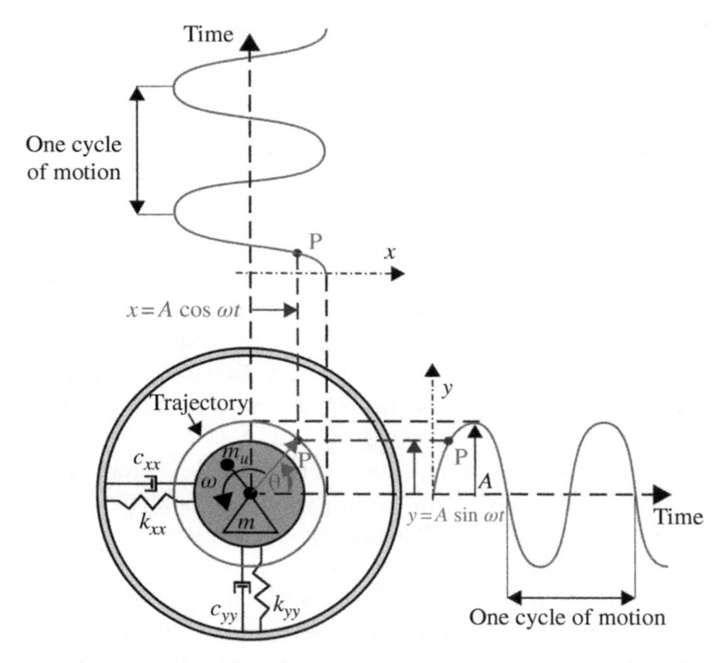

Figure 2.17 Combination of oscillation in two directions taking into account the main stiffness and damping coefficients of the bearing.

at any point in time from the relationship $\Theta = \omega t$. If we consider the motion of the point P in a vertical direction (along the y-axis), then, as a function of time, the displacement can be recorded using a function expressed by Eq. (2.16).

$$y = A \sin \omega t \tag{2.16}$$

From the mathematical point of view, the trajectories of the bearing journal, i.e. the motion of the rotor journal in time, can be considered as a combination of motion of a material point, presented in the previous subsections, in two perpendicular directions. The form of oscillation depends on the values of stiffness, damping and mass coefficients. As presented above, the results also depend on the type and frequency of the excitation force.

A system with the following coefficients was used to generate Figure 2.18:

- $m = 1.5\,\text{kg}$
- $c_{yy}, c_{xx} = 6400\ \text{N·s/m}$
- $k_{yy}, k_{xx} = 6.63 \cdot 10^6\ \text{N/m}$

The value of the excitation force was 6 N and its frequency was 1884 rad/s. Figure 2.18a presents oscillations in two directions and their combination in the form of trajectory. Because the dynamic coefficients of the bearings and the excitation were symmetrical, the result was a circle. With the differences between these coefficients, the trajectories may take the shape of an ellipsis. Figure 2.18b and c shows two sample results, where the damping coefficients c_{xx} and c_{yy} were doubled, respectively. As a result, trajectories were obtained in the form of an ellipsis appropriately altered.

2.3 Cross-Coupled Stiffness and Damping Coefficients

Cross-coupled stiffness and damping coefficients are associated with the introduction of additional stiffness and damping into the system. According to the previously adopted definition, the values of the coefficients k_{xy} and c_{xy} directly influence the displacements in the X direction as the force is applied in the perpendicular (Y) direction. Similarly, the coefficients k_{yx} and c_{xy} should be considered – they directly affect the displacements in the direction Y when the force is applied in the perpendicular X direction. A view of oscillation in which the main and cross-coupled coefficients are included is shown in Figure 2.19.

The influence of cross-coupled damping coefficients is presented in Figure 2.20. Figure 2.20a contains the results for the system without cross-coupled damping coefficients. In Figure 2.20b, cross-coupled damping coefficients are added in the xy direction only (as a negative value). The addition of negative damping increased the oscillation amplitude in the X direction. In Figure 2.20c, c_{yx} damping was added (there was no c_{xy} damping in the system), which resulted in a significant decrease in the oscillation amplitude in the Y direction. The observed changes are the same as in the previous case, where only one degree of freedom was considered.

The influence of cross-coupled stiffness coefficients is presented in Figure 2.21. Figure 2.21a contains the results for a system without cross-coupled stiffness coefficients. In Figure 2.21b, cross-coupled stiffness coefficients are added in the xy direction only (as a negative value). The addition of negative stiffness resulted in increased oscillation

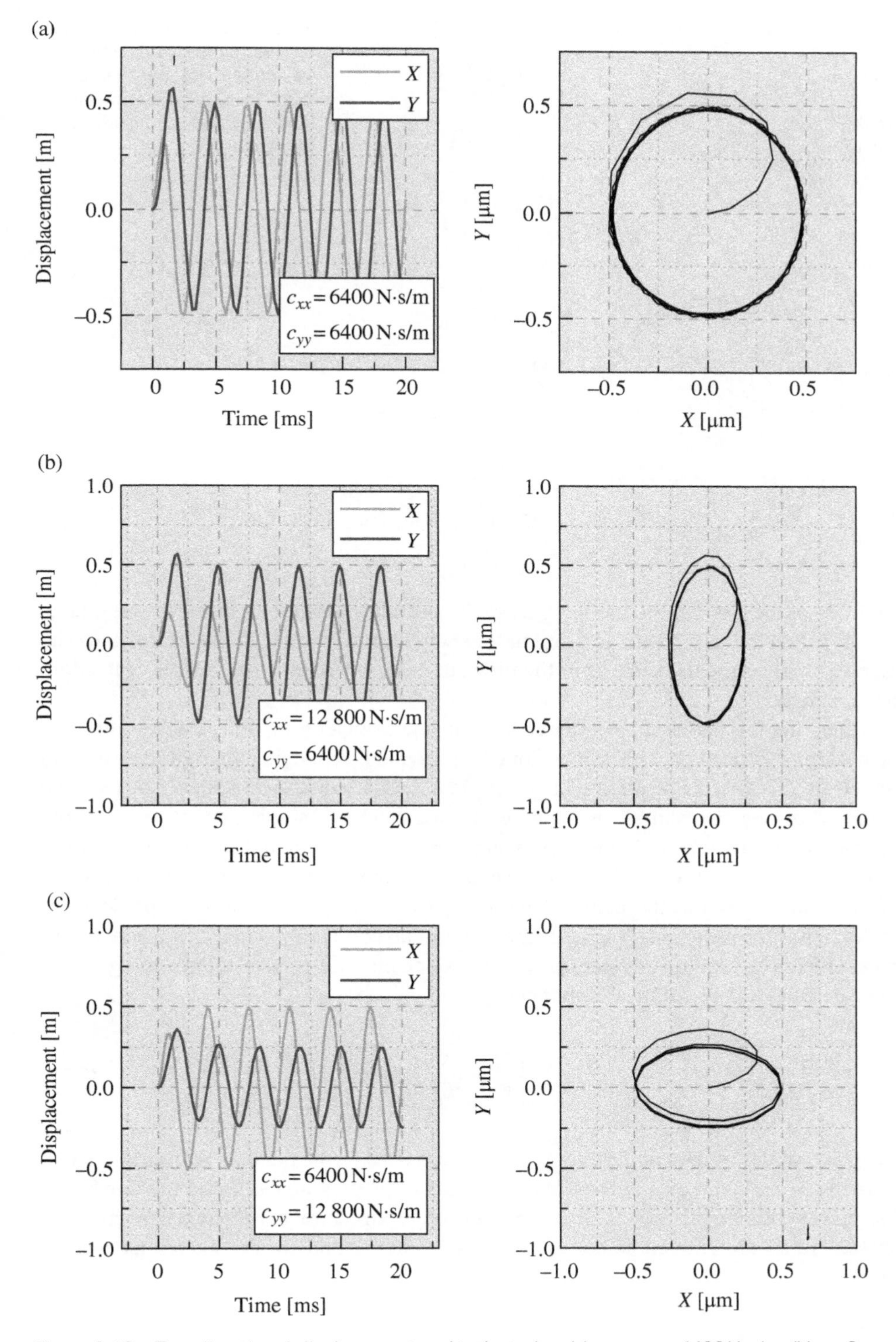

Figure 2.18 Two-directional displacement and trajectories: (a) $c_{xx} = c_{yy} = 6400\,\text{N·s/m}$; (b) $c_{xx}{\cdot}2$; and (c) $c_{yy}{\cdot}2$.

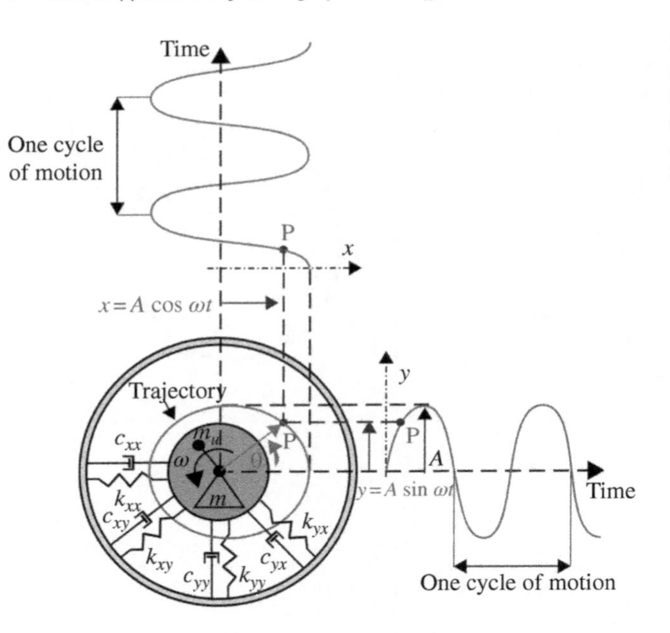

Figure 2.19 Combination of oscillation in two directions taking into account the main and cross-coupled stiffness and damping coefficients of the bearing.

amplitude in the X direction. In Figure 2.21c, k_{yx} stiffness (without c_{xy} stiffness) was added, which resulted in a significant decrease in the oscillation amplitude in the Y direction. The observed changes are the same as in the previous case, where only one degree of freedom was considered.

The fact that it is necessary to enter cross-coupled coefficients k_{xy} and c_{yx} as negative values is worth pointing out. This can be done by introducing changing appropriate values to negative in Eq. (2.4), the result of which is Eq. (2.17). This can also be achieved by entering k_{xy} and c_{xy} as negative values. In the case of cross-coupled damping coefficients c_{xy}, the minus sign does not mean that energy is supplied to the system and the system is stimulated to oscillate, but it results from the system design in the adopted reference system. If a positive value is used at the point where a negative value of the damping coefficient is entered, it has the same meaning as the negative value in other systems. The negative stiffness coefficient k_{xy} should be interpreted in the same manner. As a result of adding it (as a negative value) higher stiffness is obtained.

$$\begin{bmatrix} m_{xx} & m_{xy} \\ m_{yx} & m_{yy} \end{bmatrix} \ddot{x}(t) + \begin{bmatrix} c_{xx} & -c_{xy} \\ c_{yx} & c_{yy} \end{bmatrix} \dot{x}(t) + \begin{bmatrix} k_{xx} & -k_{xy} \\ k_{yx} & k_{yy} \end{bmatrix} x(t) = \begin{bmatrix} F_x(t) \\ F_y(t) \end{bmatrix} \tag{2.17}$$

Negative values of the coefficients k_{xy} and c_{xy} are used in the adopted reference system, where the x-axis is directed to the right, the y-axis is directed upwards and the unbalance mass moves counterclockwise. In other reference systems, a frequently used convention is that k_{yx} and c_{yx} are entered as negative values (while k_{xy} and c_{xy} are entered as positive values).

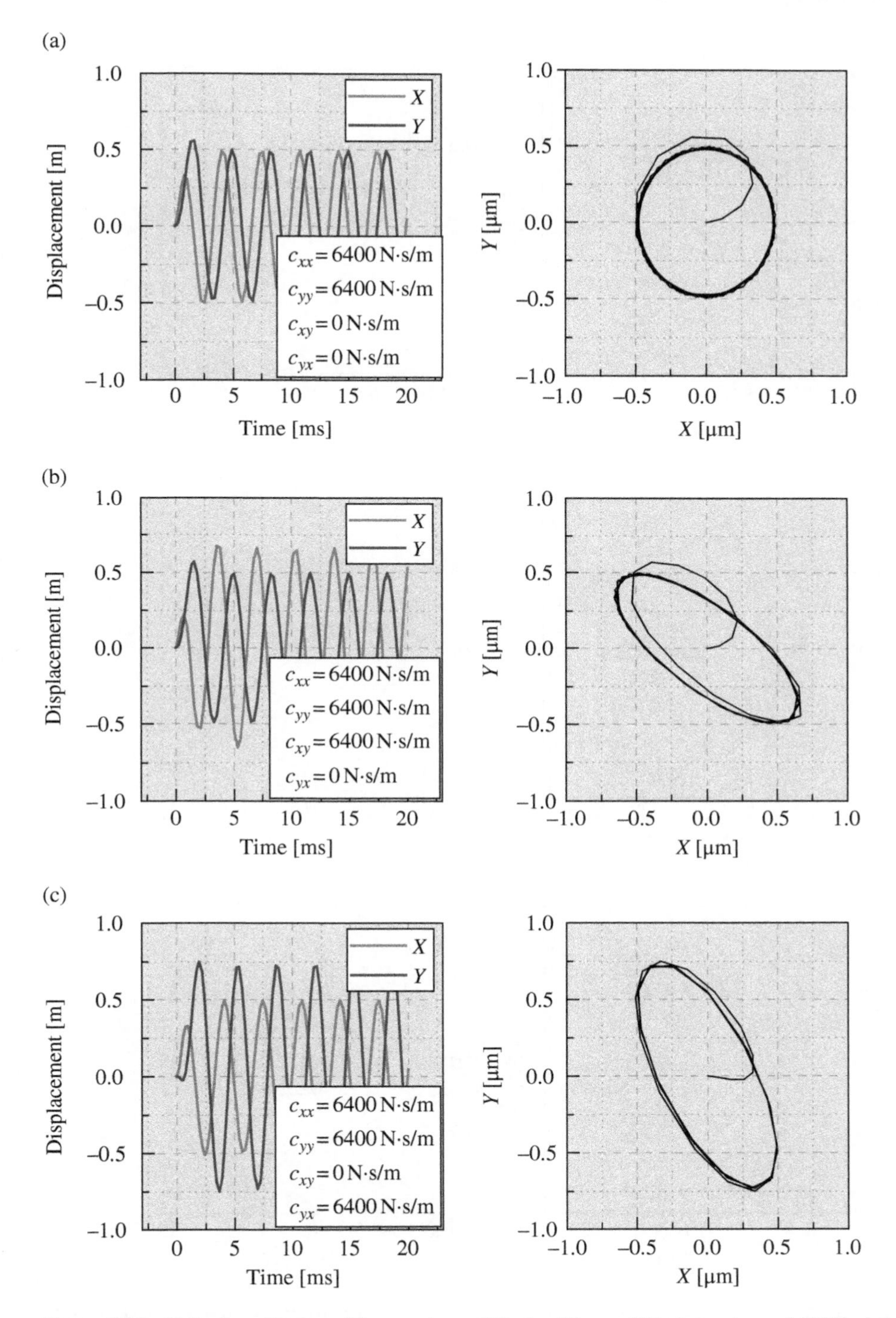

Figure 2.20 Trajectory changes: (a) c_{xy} and c_{yx} = 0 N·s/m; (b) c_{xy} = 0 N·s/m and c_{yx} = 6400 N·s/m; and (c) c_{xy} = 0 N·s/m and c_{yx} = 6400 N·s/m.

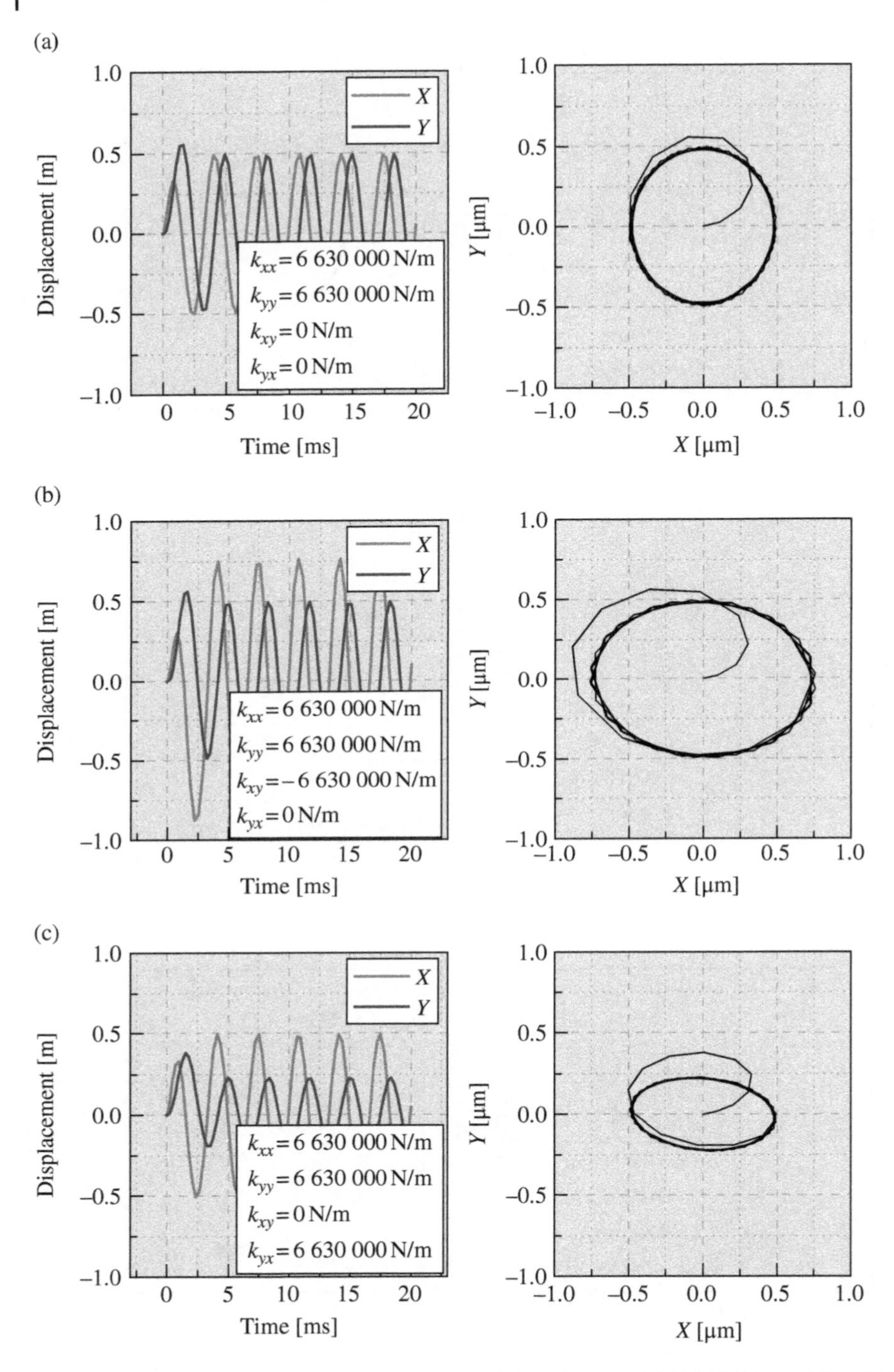

Figure 2.21 Trajectory changes: (a) k_{xy} and k_{yx} = 0 N/m; (b) k_{xy} = 6 630 000 N/m and k_{yx} = 0 N/m; and (c) k_{xy} = 0 N/m and k_{yx} = 6 630 000 N/m.

2.4 Summary

As part of the adopted simplifications, the bearing was treated as a movement of mass in two planes. Of course, the bearing itself is a much more complex object, and its movement is also influenced by external factors such as weight distribution of the rotor, sealing, or the influence of the coupling and the drive. The conditions of the support can also largely determine the dynamic properties of the rotor. Often in the initial stages of calculations it is assumed that the support is rigid and does not affect the operation of the bearing.

Different types of bearings have certain characteristics. If in the mechanical system the stiffness k_{xx} and k_{yy} are different, we are dealing with anisotropic properties in horizontal and vertical directions. Damping may also have anisotropic properties. In the case of a normal elastic mechanical system, the diagonal damping coefficients are equal, i.e. $c_{xy} = c_{yx}$. In the case of plain bearings these are almost always different, i.e. $c_{xy} \neq c_{yx}$. It can be found that a characteristic feature of hydrodynamic plain bearings is the unevenness of "cross-coupled" ratios of reaction and displacement.

On the basis of some relationships related to stiffness and damping coefficients of bearings, it is possible to draw conclusions on the stability of the system (e.g. when changing the value of cross-coupled damping coefficients to negative when increasing the rotational speed). The examples described in this chapter serve as a visualization of the influence of dynamic coefficients of bearings on the trajectories generated by them without showing the characteristic features of different types of bearings.

The analysis showing the influence of various types of excitation force on the movement of a system with one or two degrees of freedom was presented. It turns out that, depending on the type of excitation force, different effects were obtained. With a sinusoidal excitation force variable in time (corresponding to rotor unbalance), the effect of mass, stiffness and damping coefficients was shown. When the damping coefficients were increased, the resonance amplitude value decreased. When the stiffness coefficients increased, the resonance curve moved toward higher values of frequencies and amplitudes. While increasing the values of mass coefficients, the resonance curve moved toward lower values of frequencies and amplitudes.

3

Characteristics of the Research Subject

Experimental tests were carried out on the laboratory test rig manufactured by SpectraQuest, Inc. The applied numerical models reflect the geometry and parameters of this station.

3.1 Basic Technical Data of the Laboratory Test Rig

The laboratory test rig for testing small-size rotors was built to analyze the rotor–bearing system and to examine defects such as bearing damage, rotor unbalance, axial misalignment, etc. (www.spectraquest.com). A photograph of the station is shown in Figure 3.1. The weight of the stand without supporting structure is approximately 60 kg.

Figure 3.2 shows a diagram of the laboratory test rig with the dimensions and the most important elements of the structure marked. The length of the station is 125 cm, its width is 36 cm, and its height is 65 cm. The coordinate system used in experimental research is also marked. The base of the laboratory test rig is a 13 mm thick steel plate. Attached to it are two C-channels with rubber supports that allow for height adjustment and leveling. The rotor is mounted in two bearing supports, but it is also possible to mount a third support. The system was driven by a three-phase motor manufactured by Marathon Electric (www.marathongenerators.com) with a maximum speed of 3450 rpm, which was regulated by a 1.5 kW SV015iC5-1F inverter manufactured by LS (www.ls.inverterdrive.com). The motor was connected to a gearbox increasing the rotational speed approximately 3.5 times. The drive system described above enables rotational speed of up to 12 000 rpm. The transmission is connected to the rotor by a rigid coupling. The diameter of the coupling is 50 mm and its length is 60 mm. The bearing supply system was driven by the Shurflo 8050-305-526 pump (www.shurflo.com). The maximum pressure generated by the pump is 35 PSI (0.24 MPa). The oil pressure during experimental tests was 23 PSI (0.16 MPa).

The tested rotor has a length, $L_{rotor} = 920$ mm. The distance from the coupling to the first support was 170 mm. The rotor was placed in two bearing supports. The distance between supports (i.e. value $l_1 + l_2$) was 580 mm. The bearing closer to the motor is marked as number 1, and the bearing on the other end of the rotor is marked as number 2, consistently throughout operation. Exactly in the middle between the supports there was a rotor disk,

Bearing Dynamic Coefficients in Rotordynamics: Computation Methods and Practical Applications,
First Edition. Łukasz Breńkacz.

Companion website: www.wiley.com/go/brenkacz/bearingdynamiccoefficients

Figure 3.1 Photograph of the laboratory test rig.

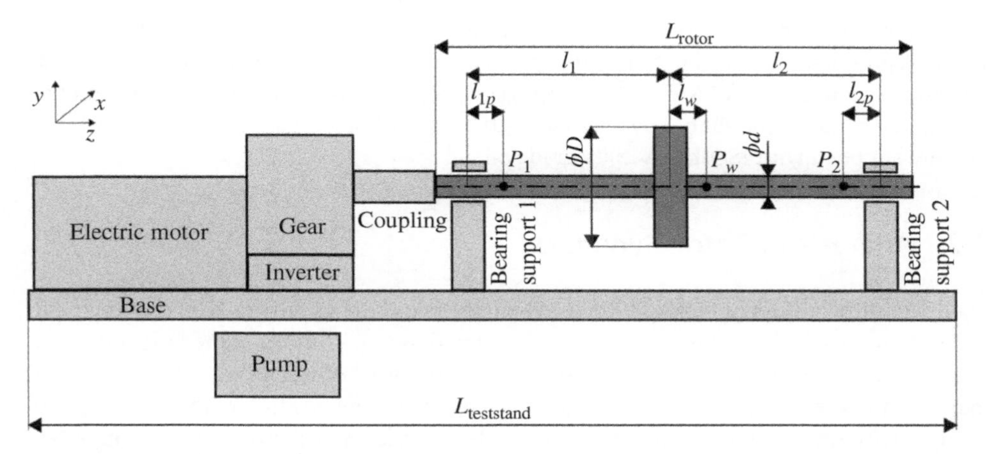

Figure 3.2 Diagram of the laboratory test rig.

thus the lengths l_1 and l_2 are equal (Figure 3.2). The diameter of the rotor was 19.05 mm (approximately ¾″), while the diameter of the rotor disk was 152.4 mm (6″). Single-axis displacement sensors are located in bearing supports. Eddy current sensors were located at points P_1 and P_2 at equal distances ($l_{1p} = l_{2p} = 25$ mm) from bearing centers. Sensors measuring displacements in the X direction were at a distance of 20 mm from the center of the bearings, while sensors measuring displacements in a direction perpendicular to the x-axis were located at a distance of 30 mm from the center of the bearings. Such a shift is necessary because eddy current sensors cannot operate in one plane. Eddy current sensors are placed at an angle of 90° to each other (and 45° to the global reference system). Excitation by means of an impact hammer took place at point P_w located at the distance $L_w = 30$ mm

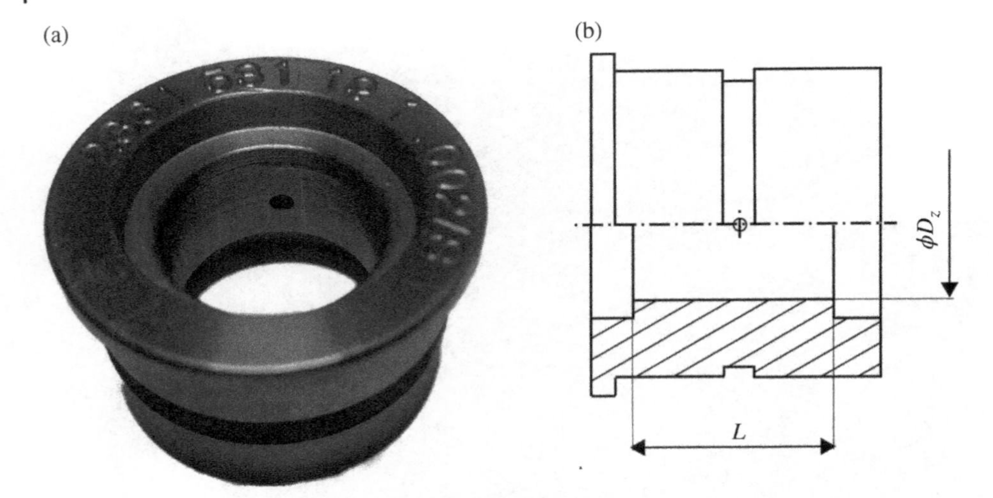

Figure 3.3 (a) Hydrodynamic bearing and (b) bearing diagram.

from the center of the rotor disk. For safety reasons, the rotor–bearing system is equipped with a lockable cover made of strong, transparent plastic.

The rotor with a diameter of 19.05 mm was supported by two hydrodynamic bearings with identical geometrical parameters. The lubrication gap related to the shaft radius was 76 μm. A photograph of the bearing housing and its schematic model is shown in Figure 3.3. The bearing length, $L = 12.6$ mm. The bearings are supplied using two holes located on the right and left side of the shaft. The diameter of the supply holes is 0.1″ (2.54 mm). The bearings are supplied with ISO 13 viscosity grade oil.

3.2 Analysis of Rotor Dynamics

Figure 3.4 shows a signal registered by eddy current sensors measuring displacement near support no. 2 (point P_2 in Figure 3.2). The diagram shows the run-up of the laboratory test rig, i.e. steady increase of the rotational speed. The presented run-up lasted 62.4 seconds. At that time the speed was increased from 1000 to 8400 rpm. The position of the rotor journal changes with the increase in rotational speed. The rotor journal changes its position, moving along the so-called semicircle of static equilibrium (Kiciński 1994).

The "traditional" course of displacement of the journal, showing how the vibration amplitude changes (after subtracting from the minimum value of the maximum value), is presented in Figure 3.5. Mean values of maximum vibration displacements of bearing journals together with their standard deviations for the entire rotational speed range are shown in Figure 3.5. The markings in the diagram are consistent with the global coordinate system shown in Figure 3.2. The increase in vibration amplitude of the second bearing from 2750 to 4500 rpm, followed by its decrease at a speed of 4750 rpm, indicates the resonance of the rotor. The higher value of the standard deviation indicated a greater difference between the maximum displacements recorded for a given rotational speed.

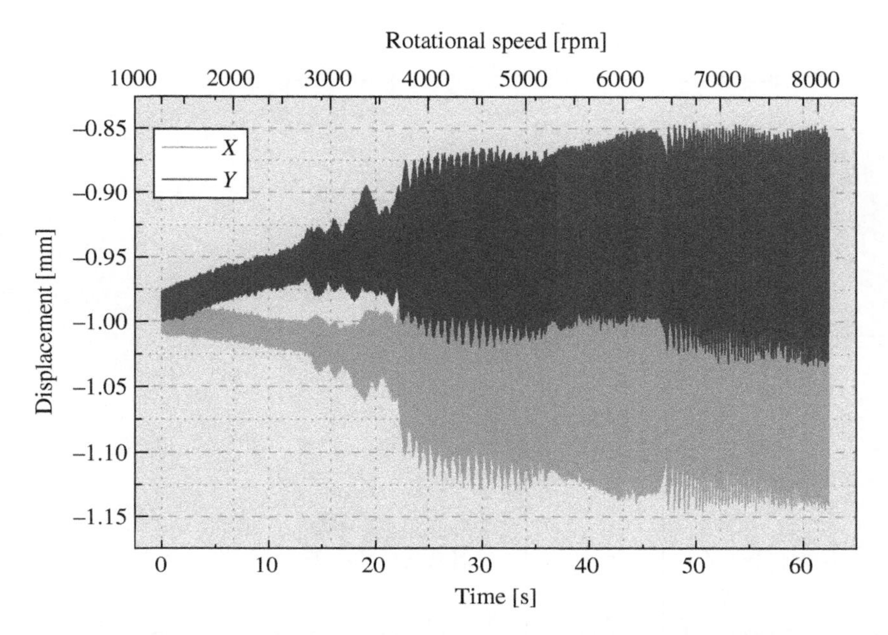

Figure 3.4 Run-up of the laboratory test rig; during 66.4 seconds the speed was increased linearly from 1000 to 8400 rpm.

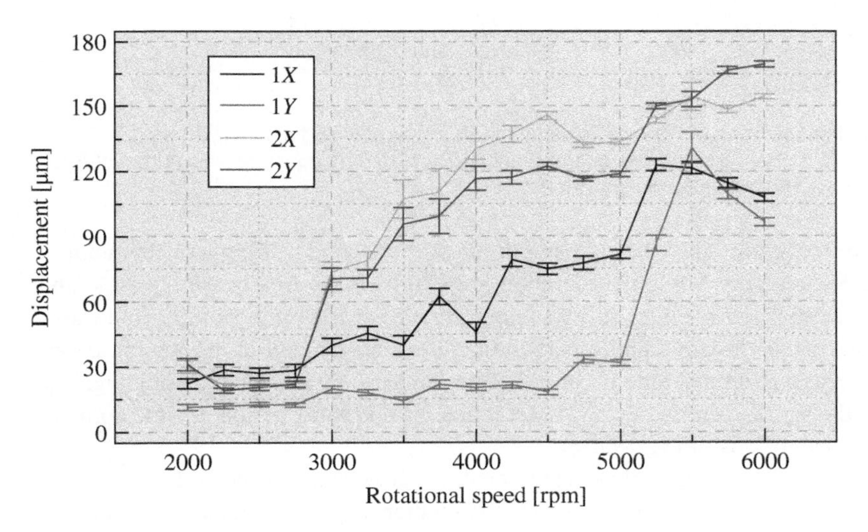

Figure 3.5 Maximum displacement of the rotor journals.

Figure 3.6 shows the vibration amplitudes recorded during stable operation of the rotor at a speed of 3250 rpm. This is a summary showing the signals recorded at the same time by four eddy current sensors located at points P_1 and P_2 (Figure 3.2). Despite the fact that two bearings with the same geometrical parameters were used in the arrangement and the rotor was almost symmetrical, the amplitudes of vibrations of the first and second bearings

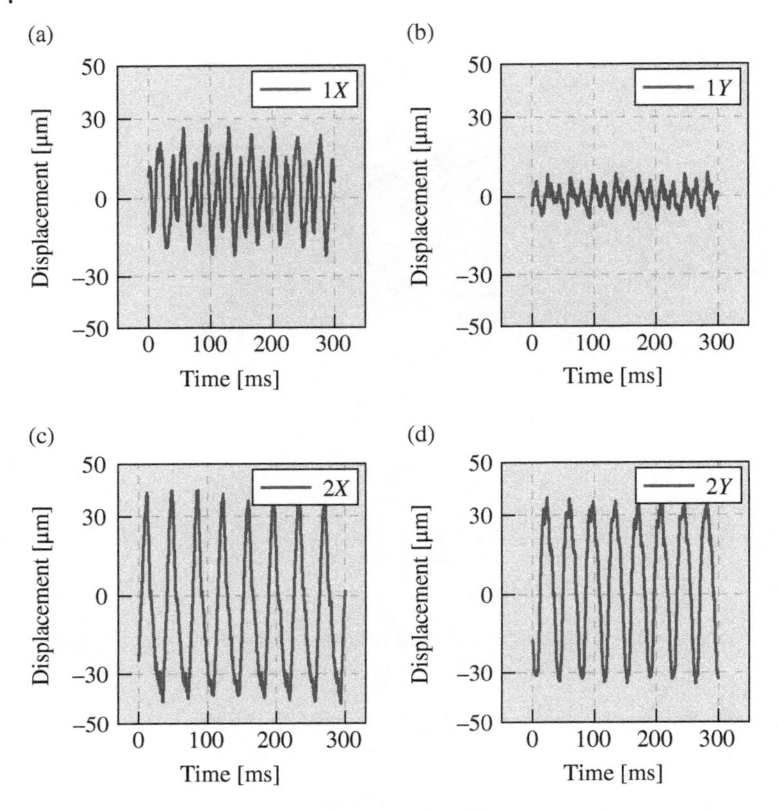

Figure 3.6 Stable operation of bearings no. 1 (a and b) and no. 2 (c and d) at 3250 rpm; the diagrams show the signal recorded in X (a and c) and Y (b and d) directions; the short time interval (0.3 seconds) shows approximately 8 revolutions.

were noticeably different. The vibration amplitude of the second bearing was about 80 μm in the X and Y directions, while the vibration amplitude of the first bearing was 50 and 10 μm, respectively. The reason for this was the influence of the rigid coupling and transmission on bearing no. 1. Although it clearly influences the dynamics of the laboratory test rig, almost every rotating machine is connected to the drive by means of a coupling. The graph shows the signal recorded during 0.3 seconds. In this short period of time approximately 8 rotor revolutions can be seen.

If a signal for a longer period of time is registered, e.g. 10 seconds (as shown in Figure 3.7), then in addition to the displacements related to the rotational speed changing 3250 times per minute, "rotor flow" can also be observed. These are changes characteristic of hydrodynamic bearings, when in addition to rotational motion, the rotor also performs a circular motion in relation to the bearing housing. The vibrations generated during stable operation are very complex. In addition to speed-related (synchronous) vibrations, the signal also contains supersynchronous and subsynchronous components and the influence of vibrations of the supporting structure. It is also visible a noise (Batko and Przysucha 2014; Przysucha et al. 2015).

In order to obtain more information on the dynamics of the system, it was necessary to carry out a fast Fourier transform (FFT) (Lyons 2010). It is an algorithm which makes it

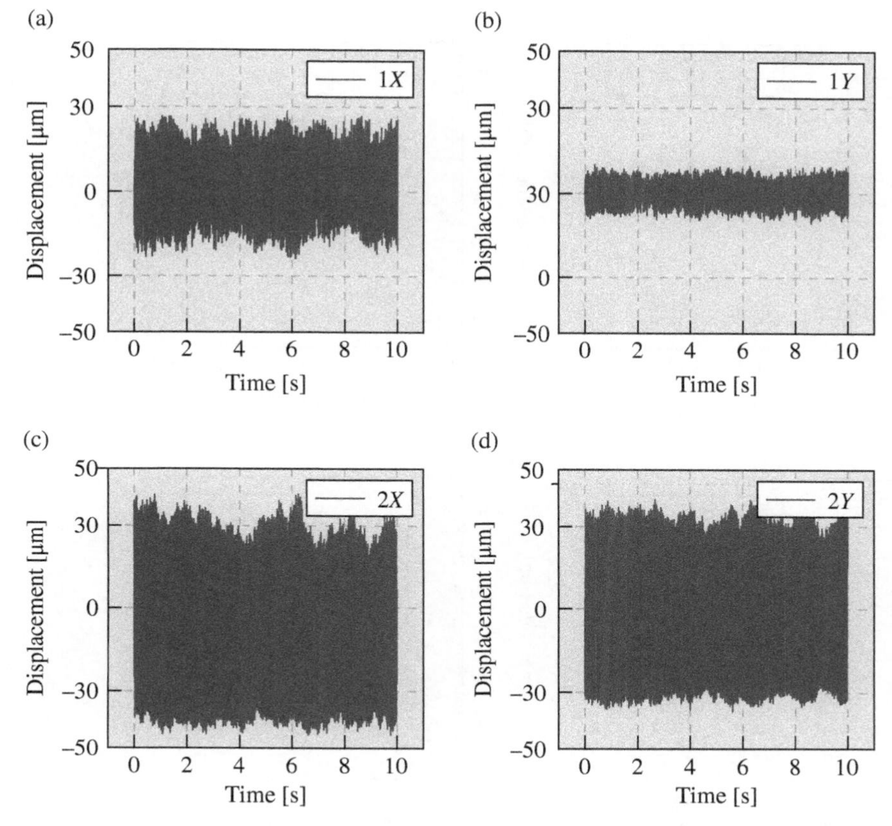

Figure 3.7 Stable operation of bearings no. 1 (a and b) and no. 2 (c and d) at 3250 rpm; the diagrams show the signal recorded in X (a and c) and Y (b and d) directions during 10 seconds.

possible to represent a signal from the time domain in the frequency domain. The analysis was used to interpret the obtained results and to calculate the dynamic coefficients of bearings. If we assume that x_0, x_{N-1} are complex numbers, then the definition of a discrete Fourier transform (DFT) can be written using the formula (3.1):

$$X_k = \sum_{n=0}^{N-1} x_n e^{-\frac{2\pi i}{N}nk}, \quad k = 0,\ldots,N-1 \tag{3.1}$$

In Matlab it is possible to implement an FFT algorithm in the form of a DFT, which is a transformation based on a discrete signal. Appendix B presents a fragment of code in Matlab software showcasing this operation. FFT diagrams of time histories from Figure 3.7 are shown in Figure 3.8. The FFT spectrum is different in two bearings and two directions. Figure 3.8a and b shows signals from the first bearing in the X and Y directions, respectively. The X and Y direction is consistent with the coordinate system designation in Figure 3.2. Similarly, Figure 3.8c and d shows data from the second bearing in the X and Y directions, respectively. The synchronous component associated with rotational speed is visible as the amplitude increase in the FFT diagram (peak at approximately 54.2 Hz). In

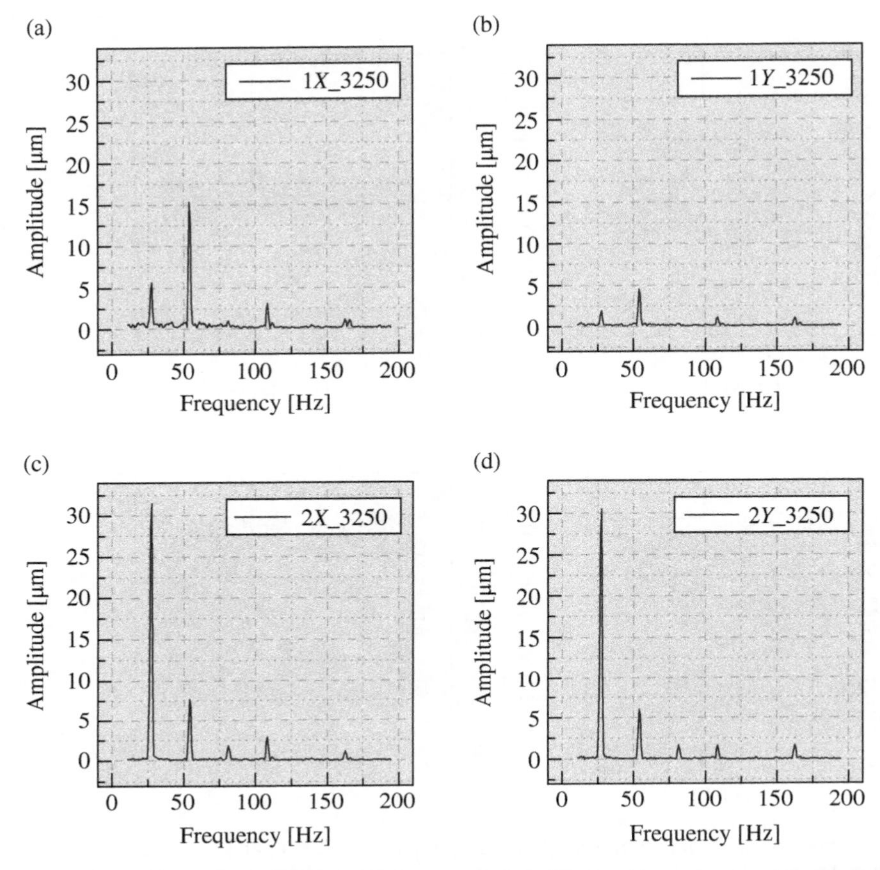

Figure 3.8 FFT analysis of the stable operation signal from eddy current sensors bearings no. 1 (a and b) and no. 2 (c and d) at 3250 rpm; the diagrams show the signal measured in *X* (a and c) and *Y* (b and d) directions.

the second bearing, the "dominant" (highest) component in the frequency domain are the subsynchronous vibrations (1/2x). FFT diagrams for rotational speeds from 2250 to 6000 rpm (in steps of 2250 rpm) are included in Appendix A.

Different vibration trajectories were observed for different rotational speeds. In order to display the results correctly, it was necessary to filter the signal using a band-pass filter. A fragment of the code prepared for this purpose in the Matlab software is presented in Appendix B. Filter parameters have been selected so that the signal of frequencies over 3× (frequency three times higher than synchronous speed) and below 1/3× (1/3 synchronous frequency) are "cut off." A filter with different parameters was used for each speed.

Trajectory of the movement of the rotor journal in bearing no. 2, representing 12 rotor revolutions, is shown in Figure 3.9. It consists of the same signal shown in Figure 3.6 in *X* and *Y* directions. Figure 3.9a contains an unfiltered signal and Figure 3.9b shows a signal filtered according to the previously described scheme.

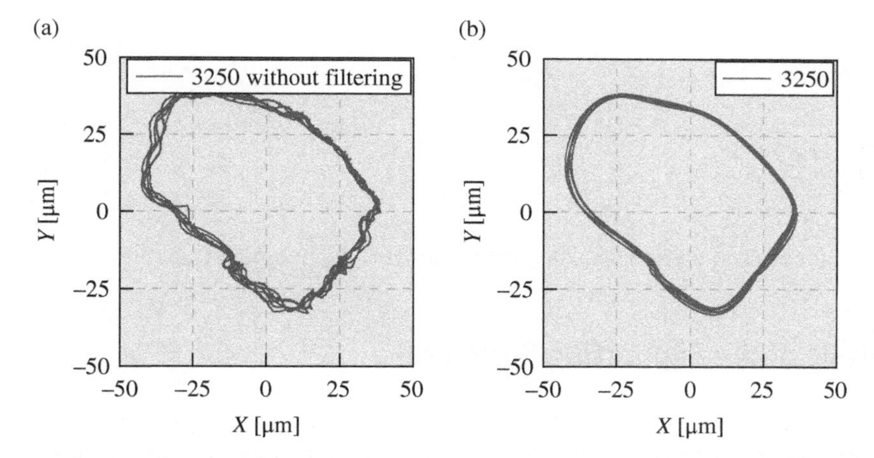

Figure 3.9 Trajectory of bearing journal movement no. 2 during stable operation at a speed of 3250 rpm: (a) signal measured by sensors and (b) signal after filtering operation.

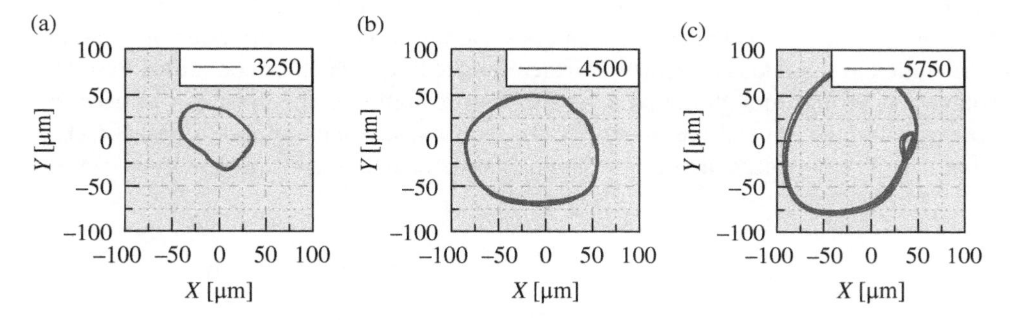

Figure 3.10 Combination of the vibration trajectories of a second bearing operating on a laboratory test rig for (a) below resonance speed (3250 rpm), (b) close to resonance speed (4500 rpm), and (c) above resonance speed (5750 rpm).

Figure 3.10 shows the operation of the rotor at three rotational speeds: below resonance speed (2750 rpm), close to resonance speed (4500 rpm), and above resonance speed (5000 rpm). A set of 16 trajectories for rotational speeds from 2250 to 6000 rpm (in steps of 2250 rpm) for two bearings are included in Appendix A.

One of the parameters for evaluating the dynamic state of rotating machinery is the observation of changes in vibration amplitude (maximum displacements) as a function of rotational speed. For this purpose, signals from four eddy current sensors were used. The diagram presented in Figure 3.5 is based on the signal generated during operation at constant rotational speeds. The calculations were made in three steps, the first for each rotational speed recorded signals lasting 10 seconds. In the second step, each of these signals is divided into 40 parts containing 12 800 measurement samples each. In the third step, the maximum displacement is calculated for each part (the minimum displacement value is subtracted from the maximum value). The mean value (3.2) and standard deviation (3.3) (Kotulski and Szczepiński 2004) were calculated from the 40 values obtained (for each speed) according to the following relationships:

$$\bar{x} = \frac{1}{n}\sum_{i=1}^{n} x_i \tag{3.2}$$

$$\sigma = \sqrt{\frac{\sum_{i=1}^{n}\left(x_i - \bar{x}\right)^2}{n}} \tag{3.3}$$

where x_i stands for the successive calculated values and n is the number of samples.

3.3 Analysis of the Supporting Structure

The vibrations of the rotating system are influenced not only by the rotor and bearings, but also by the entire supporting structure (Cannon 1973). Figure 3.11 presents a cascade diagram based on the data collected by means of an accelerometer placed on the second bearing support in the *X* direction (horizontal). A rotational speed range from 1000 to 8400 rpm was recorded. The speed is indicated on the oblique axis. The horizontal axis indicates the frequency of the signal after the FFT analysis. On the vertical axis the value of vibration acceleration amplitude is presented. Further oblique lines, which can be routed from the bottom left corner through the peaks of increased vibration amplitudes, can be used for synchronous (1×), supersynchronous (2×, 3×, etc.) and subsynchronous (1/2×, 1/3×, etc.) vibrations. The horizontal lines of the graph considered individually present the results of

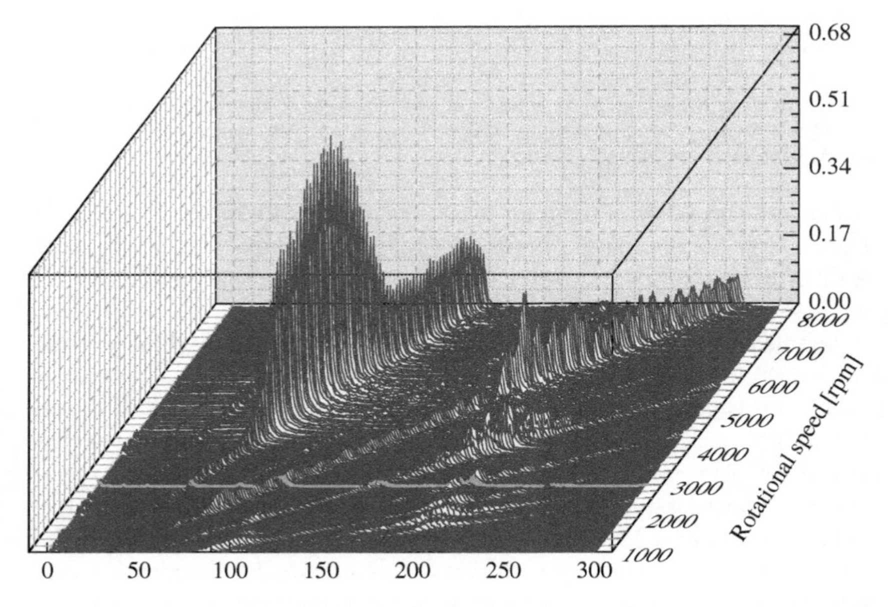

Figure 3.11 Cascade diagram based on the accelerometer signal placed on support no. 2 in the *X* direction (horizontal); on the drawing is highlighted the result of the FFT analysis at a rotational speed of 3000 rpm (dark gray horizontal line).

the Fourier signal analysis for subsequent rotational speeds. The dark gray horizontal line shows the results of the FFT analysis of a signal from a sensor placed on a bearing support at a speed of 3000 rpm. Cascade diagrams for two bearings, in two directions, are shown in Appendix A.

In order to identify the dynamic properties of the supporting structure (Uhl 1997) of the tested laboratory test rig, a modal analysis was carried out. It is a technique for testing the dynamic properties of mechanical objects. The eigenfrequencies, their forms and damping values are determined (Maia et al. 1998). Modal analysis allows to determine the correctness of the equipment's operation, reduce the risk of sudden failure, and reduce the level of generated noise. It is possible to compare the frequency of excitation in the system with the eigenfrequencies. The overlapping of these values leads to resonance (Uhl 2006, 2008; Uhl and Lisowski 1996). Five triaxial accelerometers from PCB Piezotronics were used in the research. Wax was used to mount the vibration acceleration sensors. A PCB impact hammer was used for excitation (Figure 4.4). In the modal analysis, a white plastic tip was used, which allows to induce the system in the frequency range up to 3 kHz. A list of the eigenfrequencies of the first 10 forms of natural vibrations and the corresponding damping coefficients are presented in Table 3.1. The higher the value of the damping coefficients, the more the individual forms of the eigenfrequencies will be dampened.

Figure 3.12 shows the first form of natural vibrations occurring at a frequency of approximately 287 Hz. The main components involved in vibration are the transverse vibrations of the bearing supports. Valuable information resulting from the analysis is the fact that the frequency of the first form of natural vibration has a high value, and therefore does not significantly affect the calculation of the dynamic coefficients of bearings.

Table 3.1 Successive eigenfrequencies and the corresponding damping values calculated for the laboratory test rig.

Frequency no.	Frequency (Hz)	Damping (%)
1	286.65	2.39
2	397.77	4.14
3	473.57	1.51
4	1060.20	2.06
5	1130.07	1.71
6	1292.66	1.76
7	1592.07	1.64
8	1772.66	0.96
9	1959.43	1.00
10	2326.92	1.00

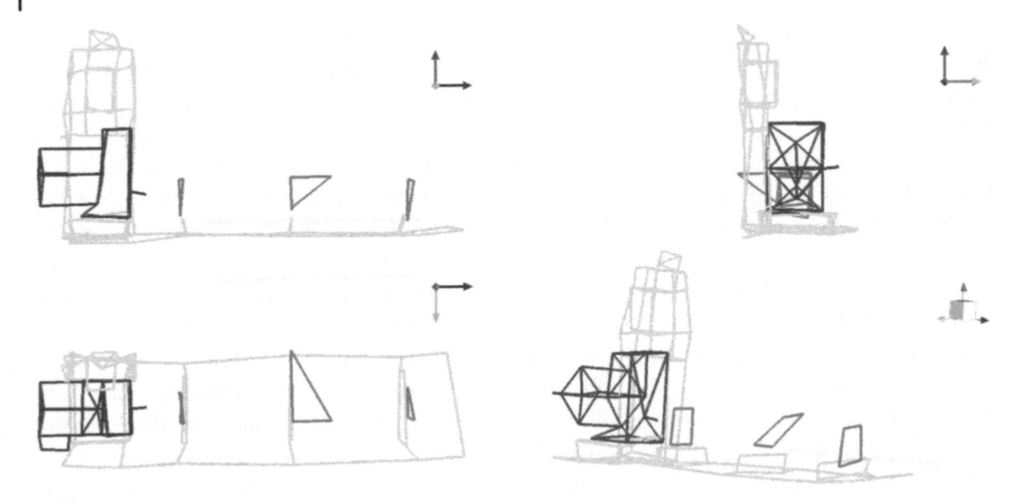

Figure 3.12 The first form of natural vibrations of the laboratory test rig.

3.4 Summary

Since experimental determination of stiffness, damping and mass coefficients of bearings requires appropriate interpretation of the experimental signal, it is necessary to know the dynamic properties of the laboratory test rig. This chapter presents a description of the laboratory test rig, detailing its most important elements and dimensions. Base characteristics of the test rig indicating the signal that can be measured during stable operation and run-up were presented.

It shows the signal recorded by eddy current sensors at a speed of 3250 rpm. On the basis of the signal measured in this way, calculations of dynamic coefficients of hydrodynamic bearings were performed. Their values are presented in Chapter 8. Particularly important in the analysis of results is the observation of the FFT signal diagrams. The operations necessary to calculate the dynamic coefficients of bearings were performed in the frequency domain. The FFT analysis is carried out for the sine waveform signal. It can be used to "decompose" this signal and determine the components of the analyzed signal (e.g. amplitude and frequency). This is a commonly used method of identifying the dynamic state of rotating machines.

The journal's vibration trajectories were created by placing vibrations generated in the X and Y directions on one diagram and filtering them appropriately. Trajectories for several rotational speeds are presented (Appendix A includes a list of trajectories recorded for two bearings) for 16 rotational speeds from 2250 to 6000 rpm (with increments of 2250 rpm). This compilation, drawn up in a single scale diagram, illustrates the behavior of the rotating machine. Making a vast generalization, it can be stated that smaller "loops" in addition to the ellipse generated by the unbalance of the rotor indicate the occurrence of subsynchronous vibrations. The curves of the main orbit of vibrations may indicate the occurrence of vibrations of higher frequencies (supersynchronous). Appendix A presents the FFT analysis of the signal recorded for the entire rotational speed range. Observation of vibration trajectories and frequency domain diagrams make the assessment of the dynamic state of the machine and its diagnostics possible (Cholewa 2014a, 2014b; Cholewa and Kiciński 1997; Komorska 2011).

Diagrams of changes (Figure 3.5) in rotor vibration amplitude as a function of rotational speed are drawn up by subtracting the minimum amplitude value of the signal registered by the eddy current sensor from the maximum value of this amplitude. Mean values and standard deviations were calculated. These values are shown on the diagram. The increase in vibration amplitude around rotational speed of 4500 rpm and its subsequent decrease is related to the rotor resonance (Lipka 1967). Larger values of the standard deviation, marked as an error in the graph, indicate greater changes in the measured amplitude and are related to lift motion.

In contrast to the results described above (Figure 3.5), the cascade diagram describes the entire supporting structure rather than the rotor vibrations. The design has a direct influence on the operation of the rotating machine. The cascade diagram, prepared on the basis of data recorded by the acceleration sensor, shows the signal for the run-up from 1000 to 8400 rpm. Figure 3.11 illustrates how the signal changes in the frequency domain during speed changes. Increasing the amplitude from the bottom left corner in oblique directions indicates vibrations changing with the rotational speed. The increase in vibration amplitude perpendicularly to the horizontal axis (not changing with the rotational speed) indicates the occurrence of resonant vibrations of the structure.

The modal analysis of supporting structure made it possible to determine the form of natural vibrations, the frequency of natural vibrations and the corresponding damping coefficients. It turns out that the first form of natural vibrations occurs at a frequency of approximately 287 Hz. This means that the resonant vibrations of the structure caused by unbalance at synchronous speed will "emerge" during operation at 17 220 rpm. Experimental research was carried out at lower speeds, but this does not mean that this form will not be excited. The form of natural vibration can be excited by higher or lower harmonics of rotational speed. However, they should not have a big influence on the dynamics of the rotor.

All the above analyses are a prelude to further experimental and numerical research. The calculations of the dynamic coefficients of bearings described in this work consisted, in short, in excitation of a rotating rotor with an impact hammer and the appropriate processing of the measured signals. Knowledge of the basic characteristics of the rotating machine makes it possible to accurately interpret the results presented in the later chapters of this book.

4
Research Tools

As part of the work experimental research and numerical calculations were carried out. In order to prepare the work, it was necessary to use a laboratory test rig with a rotor working on hydrodynamic bearings and extensive measuring equipment was also used. This instrumentation is available at the Institute of Fluid-Flow Machinery at the Polish Academy of Sciences in Gdansk. The measurements were carried out in accordance with relevant standards (Kosmol 2010). The vibrations of non-rotating parts were measured in accordance with ISO 10816-1:1995 (1995). Measurement of rotor vibrations was compliant with ISO 10817-1:1998 (1998). The balancing of the rotor is based on ISO 1940-1:2003 (2003). Schematic drawings were made using the Inkscape vector graphics software (www.inkscape.org). The graphs were prepared in vector form using Origin 2018 Pro software (www.originlab.com).

4.1 Test Equipment

The LMS International SCADAS Mobile was used as a data acquisition module (Figure 4.1). The analyzer was equipped with 40 measurement channels. It was possible to connect two tachometers and control, e.g. inducer amplifiers (www.plm.automation.siemens.com). Shielded cables were used to connect the sensors. SCADAS Mobile was connected to a laptop containing Test.Lab 11B software via an Ethernet connection. The analyzer was used to measure signals from a impact hammer, tachometer, accelerometers, and eddy current sensors.

Four eddy current sensors CWY-DO-501A were used for measurement of rotor displacement (Figure 4.2a). The sensors were connected to the SCADAS Mobile analyzer via a demodulator (www.spectraquest.com) (Figure 4.2b). The sensitivity of eddy current sensors was approximately 4.00 mV/μm (Global Sensor Technology 2012). Their measuring range was 1 mm, the resolution was 0.1% of the measuring range, and the non-linearity was about 0.7% of the measuring range. The sensors were 6 mm in diameter and 45 mm in length.

Four single-axis accelerometers type 608A11 (PCB Piezotronics 2011) were used to measure accelerations of vibrations. The accelerometer is shown in Figure 4.3a. The sensitivity of

Bearing Dynamic Coefficients in Rotordynamics: Computation Methods and Practical Applications,
First Edition. Łukasz Breńkacz.

Companion website: www.wiley.com/go/brenkacz/bearingdynamiccoefficients

Figure 4.1 SCADAS mobile analyzer.

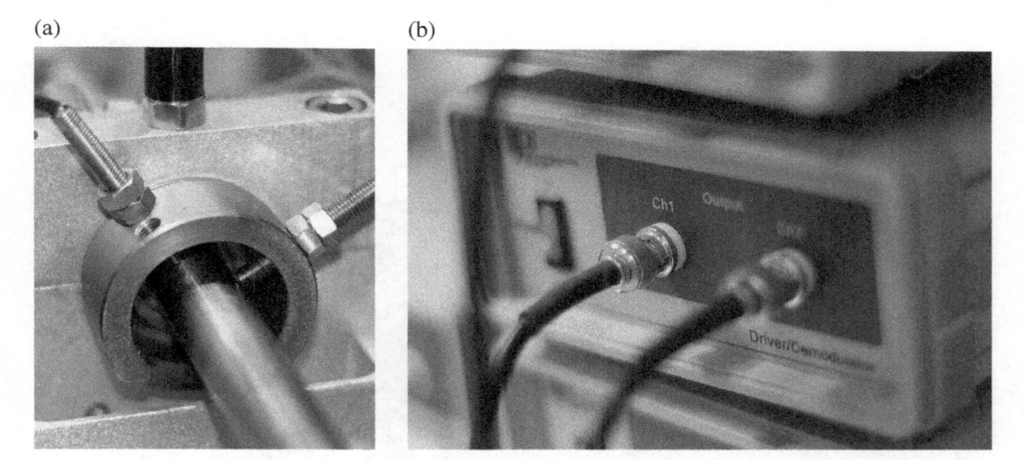

Figure 4.2 (a) Eddy current sensors placed at a 90° angle to each other and perpendicular to the rotor axis and (b) module used for processing signals from eddy current sensors – demodulator.

accelerometers ranged from 94 to 100 mV/g. The accelerometers were mounted in two bearing supports in directions perpendicular to each other and to the rotor's axis. Sensitivity of accelerometers was measured with the PCB 699A02 portable calibrator PCB 699A02 (PCB Piezotronics 2008) shown in Figure 4.3b. The calibrator operates at a frequency of 159.2 Hz, the acceleration of vibrations generated by this device is 9.81 m/s^2 (www.pcb.com).

The impact hammer is shown in Figure 4.4 (PCB Piezotronics 2007). Its sensitivity equaled 2.25 mv/N. The upper value of the measuring range declared by the manufacturer was 2224 N. The resonance frequency was greater than 22 000 Hz. The hammer weighed 0.16 kg, had a head diameter of 1.57 cm, and was 21.6 cm in length. During the measurements, a hard impact cap (ST STL) was used (the maximum force possible was 360 N).

The rotational speed was measured with the Optel Thevon 152 G7 laser tachometer (www.optel-thevon.fr). The sensor and switch are shown in Figure 4.5. The tachometer can measure up to 100 000 pulses per second. Operation at temperatures from −50 to 120 °C is

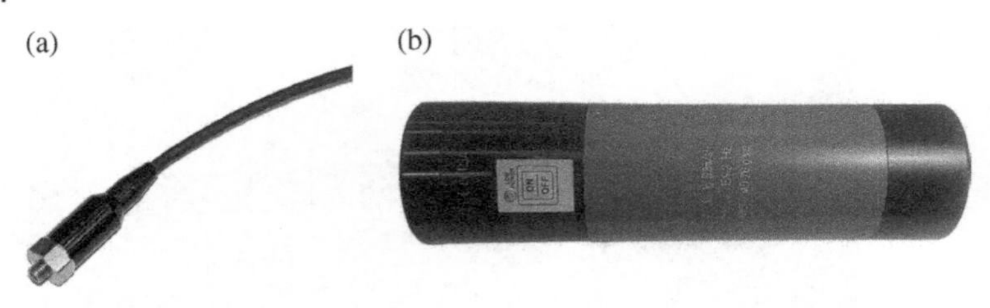

Figure 4.3 (a) Accelerometer used to measure accelerations of vibrations of the bearing supports and (b) portable accelerometer calibrator.

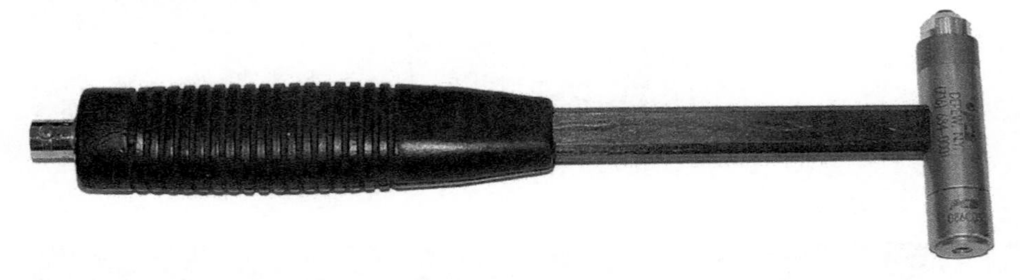

Figure 4.4 Impact hammer.

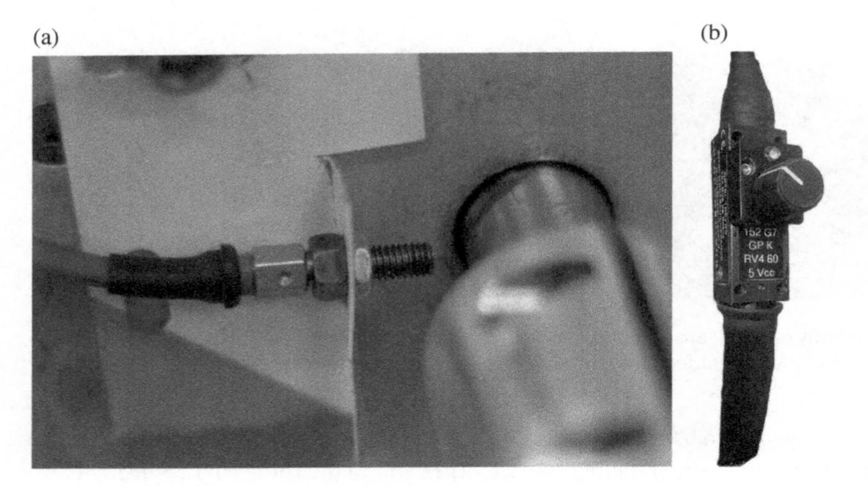

Figure 4.5 (a) Speed measurement using a laser sensor and (b) fiber optic switch.

possible. During operation, a signal was generated; the displacement increase time was less than 150 ns, while the falling time was less than 0.5 μs.

As part of the study, the analysis of hammer impact in slow motion was carried out using the Phantom v2512 high-speed camera (Figure 4.6) manufactured by Vision Research (Phantom 2015). The camera is equipped with a high-quality CMOS sensor, generating 1 MP images with a maximum resolution of 1280 × 800-1 Mpx. One pixel has a diameter of 28 μm. The camera's 25 Gpx/s bandwidth allows movies at up to one million frames per second to be recorded. At maximum resolution it is possible to record video at 25 600 frames per second. The maximum sensitivity of the camera is ISO-6400. The camera is connected to a

Figure 4.6 The Vision Research Phantom v2512 camera (right), the LED lamps used for illumination (center), and the laboratory test rig (left).

measuring computer via an Ethernet port. PCC (Phantom Camera Control) software from Vision Research (https://www.phantomhighspeed.com/news/whoweare/aboutvri) and TEMA Motion from Image Systems (www.imagesystems.se) were used for signal analysis.

Two 60 W Easy LED PRO 2X Oslon 60 LED lamps (www.easyled.fi) were used for illumination. Each lamp provided 6000 lm of illumination (Easy LED 2014). Two lenses were used for the recordings: Zeiss Plannar T* 50 mm F/1.4 ZF.2 (www.zeiss.com) and Nikon 200 mm F4.0. ED-IF AF Micro-Nikkor (www.nikon.com).

During the preparation of the laboratory test rig two instruments were used. The first was Diamond 401 – a universal vibration meter and analyzer with the function of balancing in own bearings (MBJ Electronics 2011). The device was manufactured by MBJ (www.mbj.com.pl) (Figure 4.7a). This device was used to balance the rotor. The second device was OPTALIGN smart RS manufactured by Prüftechnik (www.pruftechnik.com) (Figure 4.7b). With this device and laser sensors, the shaft supported on plain bearings and the motor shaft were axially aligned.

4.2 Test.Lab Software

Test.Lab (Figure 4.8) is used to acquire and process experimental data from the SCADAS Mobile multi-channel analyzer. It is a complete, integrated environment combining high-speed multi-channel recording with tools for data analysis and report generation. The program allows cascade charts to be created, rows to be analyzed, and to operate on the signal in the time and frequency domain. The software was created by LMS International, and has been developed by Siemens since 2013 (www.pl.automation.siemens.com). Data from measurement sensors were collected using Test.Lab version 11B.

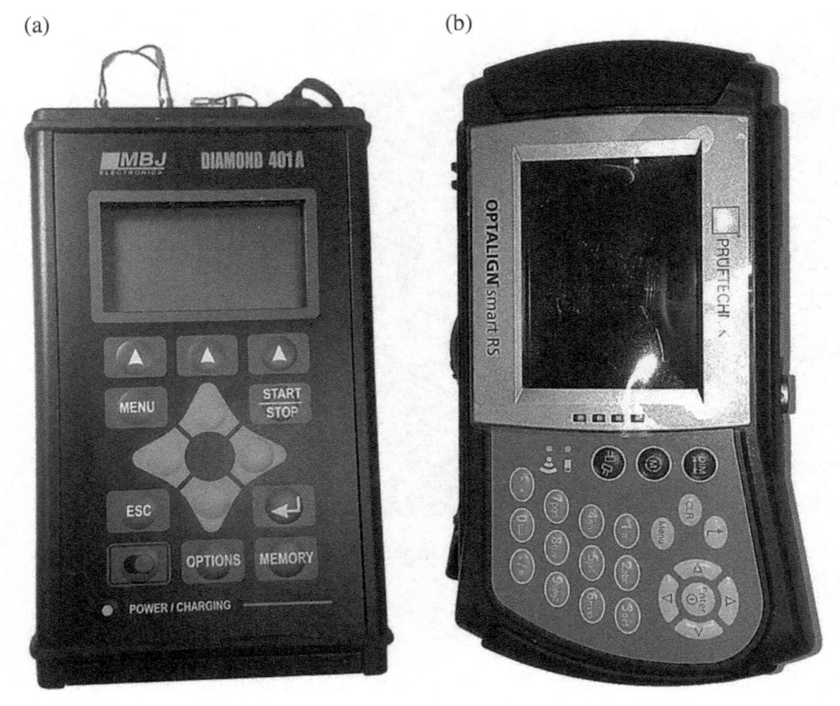

Figure 4.7 (a) MBJ Diamond 401 measuring device used for balancing the rotor and (b) the OPTALIGN Smart RS device for shaft alignment.

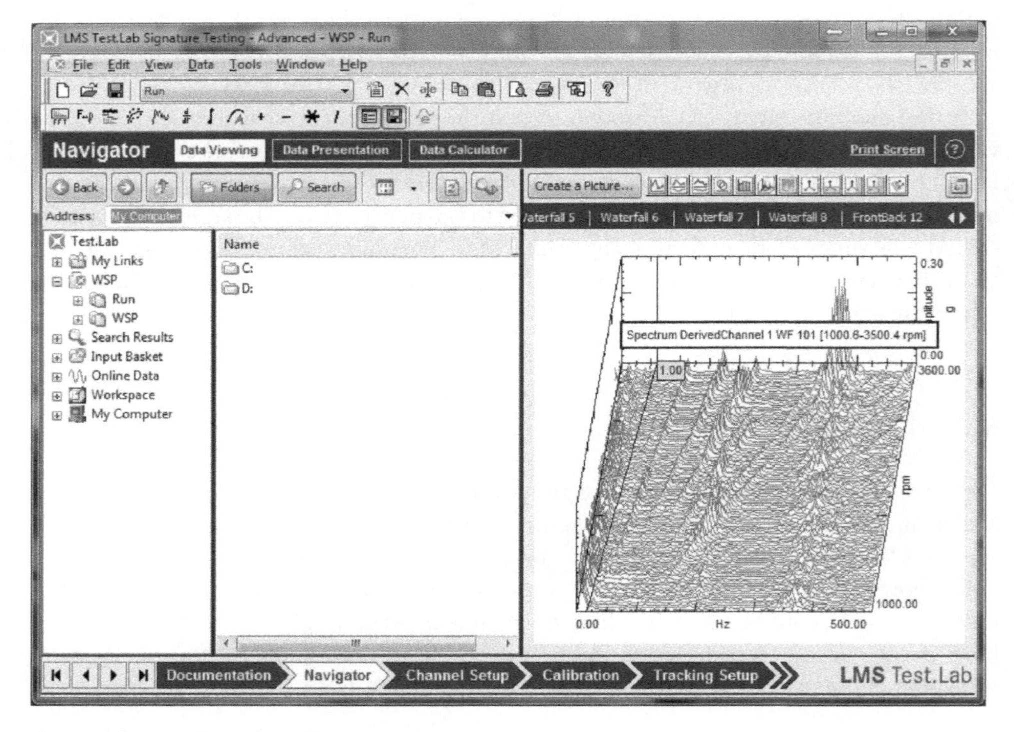

Figure 4.8 Interface of the Test.Lab 11B software.

4.3 Samcef Rotors Software

The Samcef Rotors software (Figure 4.9) enables numerical analysis of rotating machines along with bearings, gears and connecting elements (www.pl.automation.siemens.com). It facilitates the dynamic analysis of machines such as aircraft engines, turbochargers, pumps, and fans. It is based on the finite element method (Bielski 2013; Dacko et al. 1994; Grabarski and Wróbel 2008; Majchrzak and Mochnacki 2004; Milenin 2010; Radwańska 2010; Rakowski and Kacprzyk 1993). The software allows the form and frequency of natural vibrations to be calculated for complex systems such as rotor–bearing–foundation and to check their response to excitation. It is possible to include non-rotating parts (such as stators and enclosures). Samcef Rotors is part of the Samcef Field software package. Samtech, which initially developed the program, was taken over by LMS International, changing its name to LMS Samtech. The application has been developed by Siemens since 2013.

The Samcef Rotors software verified the algorithm for experimental determination of mass, damping and stiffness coefficients. The program was also used to analyze the sensitivity of the described method.

4.4 Matlab Software

Matlab (Figure 4.10) by The MathWorks, Inc. is a high-level programming language (www.mathworks.com). Its commands, operators, and functions are used for numerical calculations (including matrices and complex numbers) (Kłosowski 2011; Mrozek and Zbigniew 2010). More than 1300 Matlab functions (described in the documentation) perform numerical algorithms (Jankowski 1983), matrix operations, and many other advanced mathematical operations (Krzyżanowski 2011; Kucharski 2000).

Matlab contains a wide range of graphical functions for creating very complex diagrams of functions (Rucka and Krzysztof 2012; Szymczyk 2006). The GUIDE environment was

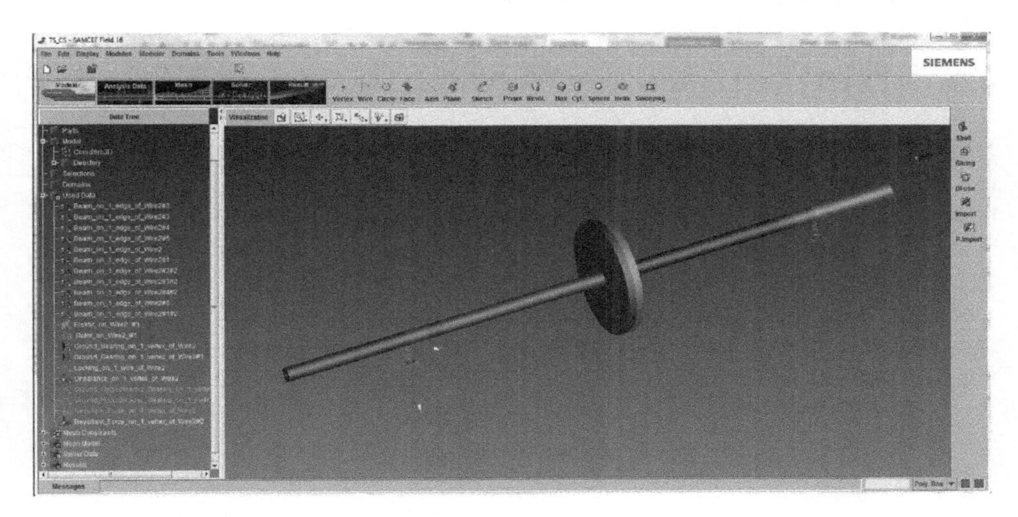

Figure 4.9 Interface of the Samcef Rotors software.

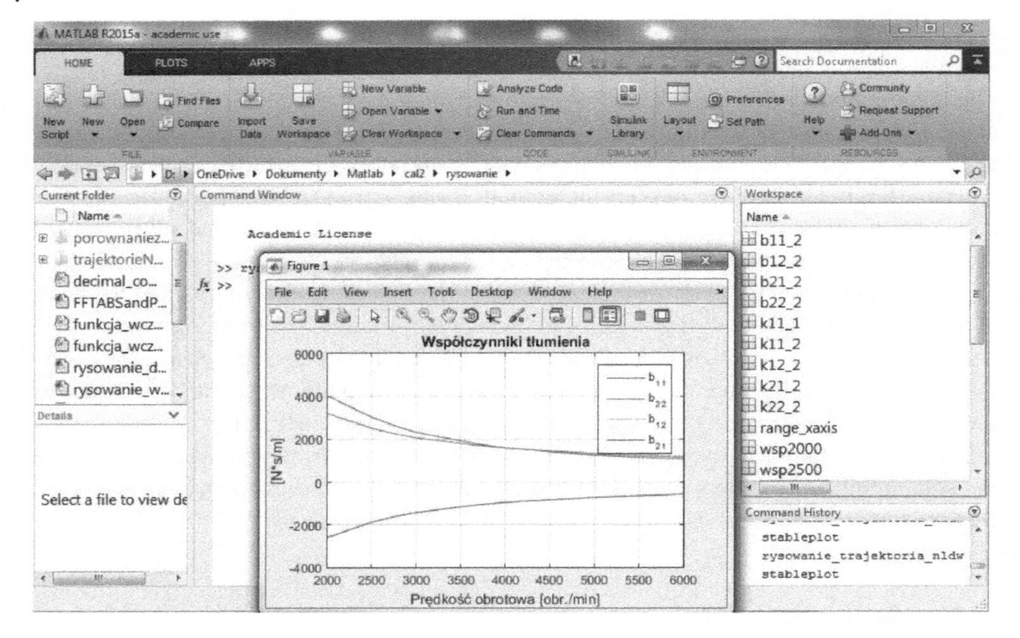

Figure 4.10 Interface of the Matlab software.

used in the development of this monograph. It makes it possible to design and implement custom graphical interfaces for the Matlab application (Lent 2013; Smith 2006).

The Matlab license was granted as part of an individual research grant at the Centre of Informatics – Tricity Academic Supercomputer and network (www.task.gda.pl). In the Matlab environment (along with the GUIDE add-on), programs for signal preparation and processing were developed.

4.5 MESWIR Series Software (KINWIR, LDW, NLDW)

MESWIR is a set of computer programs combined into a coherent environment which can be used for the analysis of rotary machines (Kiciński 2006) For many years, it has been developed at the Institute of Fluid Flow Machinery at the Polish Academy of Sciences in Gdansk in the Turbine Dynamics and Diagnostics Department (formerly Department of Dynamics of Rotors and Slide Bearings), under the management of Professor Jan Kiciński. These programs are unique and commercially unavailable. They can be used to model dynamic interactions of complex rotor–bearing–foundation systems with linear and non-linear analysis methods. It is also possible to model heat exchange in bearing nodes. An example of a model prepared with the NLDW software is shown in Figure 4.11.

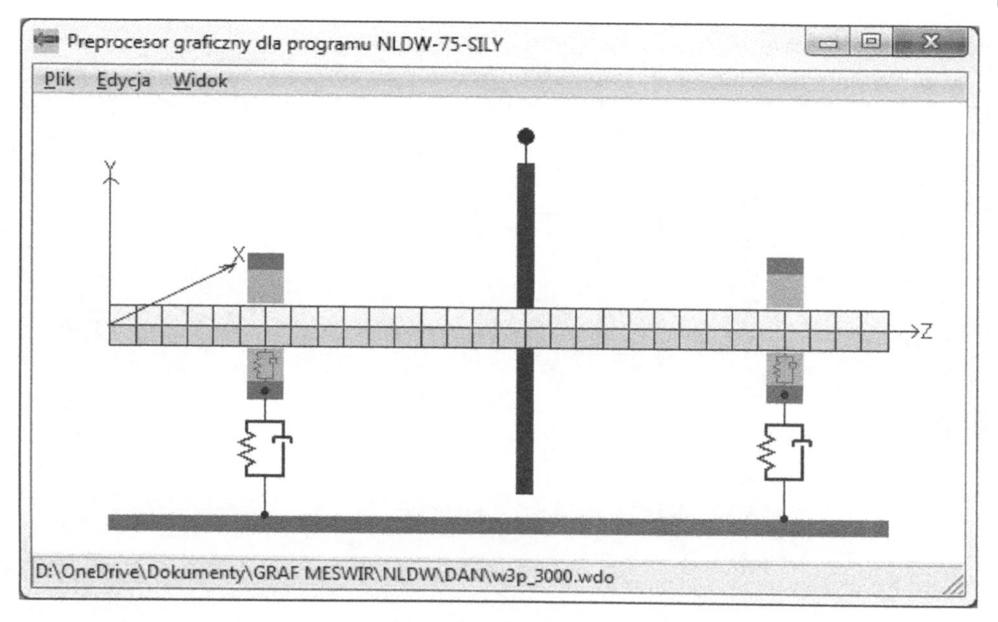

Figure 4.11 Interface of GRAFMESWIR – a graphical processor of the NLDW software.

KINWIR, LDW and NLDW programs were used together with appropriate graphic programs. Linear stiffness and damping coefficients were calculated in the KINWIR program, and the displacements of the journals were determined using the LDW program. Nonlinear calculations were carried out using the NLDW program.

4.6 Abaqus Software

The Abaqus software (Figure 4.12) is a complete, modern tool for conducting calculations in the field of continuous media and fluid mechanics (Abaqus 2015). It uses the finite element method algorithm. The Abaqus program has been developed since 1978, initially by Hibbit, Karlsson and Sorensen, and since 2005 by Dassault Systèmes (2020). Abaqus offers a graphical environment for model creation and visualization of calculation results. Abaqus/CAE is a complete pre- and post-processor for the entire calculation process (Helwany 2007).

Abaqus was used to verify the correctness of experimental calculations of stiffness and damping coefficients of hydrodynamic bearings. Analyses were performed in the Abaqus/Explicit module designed for the analysis of fast-changing dynamic phenomena. An explicit method of integration of the equations of motion was applied, so that stable solutions could be achieved even under fast-changing, extreme dynamic loads.

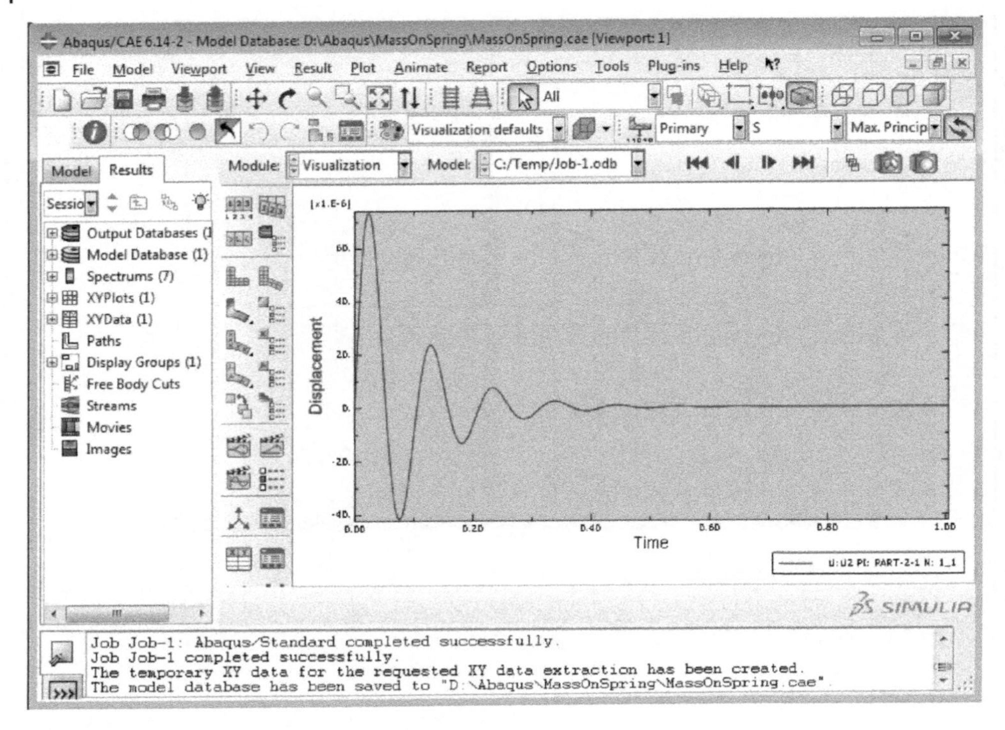

Figure 4.12 Interface of the Abaqus/CAE software 6.14-2.

5

Algorithms for the Experimental Determination of Dynamic Coefficients of Bearings

This chapter presents an algorithm for the experimental determination of dynamic coefficients of bearings. A numerical model was built in the Samcef Rotors software, and then the calculation algorithm was verified on its basis.

5.1 Development of the Calculation Algorithm

The movement of a material point can be described by Eq. (5.1). After applying Laplace transform and Euler–Lagrange equations we obtain Eqs. (5.2)–(5.4) (Breńkacz 2015a):

$$\ddot{x}+c\cdot\dot{x}+k\cdot x=f(t) \tag{5.1}$$

$$m(j\omega)^2X(s)+c(j\omega)X(s)+kX(s)=F(s) \tag{5.2}$$

$$\left[\left(k-m\omega^2\right)+d(j\omega)\right]X_0e^{j\phi_x}=F_0e^{j\phi_F} \tag{5.3}$$

$$\left[\left(k-m\omega^2\right)+d(j\omega)\right]X_0\left(\cos\phi_x+j\sin\phi_x\right)=F_0\left(\cos\phi_F+j\sin\phi_F\right) \tag{5.4}$$

In the above formulas, x stands for displacement, $\dot{x}$ for speed, and $\ddot{x}$ for acceleration. Mass is denoted by "m," stiffness as "k," excitation force as F_0, and rotational speed as ω. The equation of motion for one bearing (for the model shown in Figure 5.1b) can be described with Eq. (5.5). These equations can be written in matrix form (Bishop et al. 1972) as (5.6). The system described in such a way provides a different response to the excitation in the X and Y directions by the extortion forces F_x and F_y, respectively. The response of the system in the X direction after excitation in the Y direction is described by the xy index. The numerical model of the rotor used for verification of the calculation algorithm was created on the basis of the previously described (in Chapter 3) laboratory test rig, located in the Institute of Fluid Flow Machinery at the Polish Academy of Sciences in Gdansk, Poland.

$$-m_{xx}\omega^2X_x-m_{xy}\omega^2X_y+k_{xx}X_x+c_{xx}j\omega X_x+k_{xy}X_y+c_{xy}j\omega X_y=F_x$$

Bearing Dynamic Coefficients in Rotordynamics: Computation Methods and Practical Applications, First Edition. Łukasz Breńkacz.

Companion website: www.wiley.com/go/brenkacz/bearingdynamiccoefficients

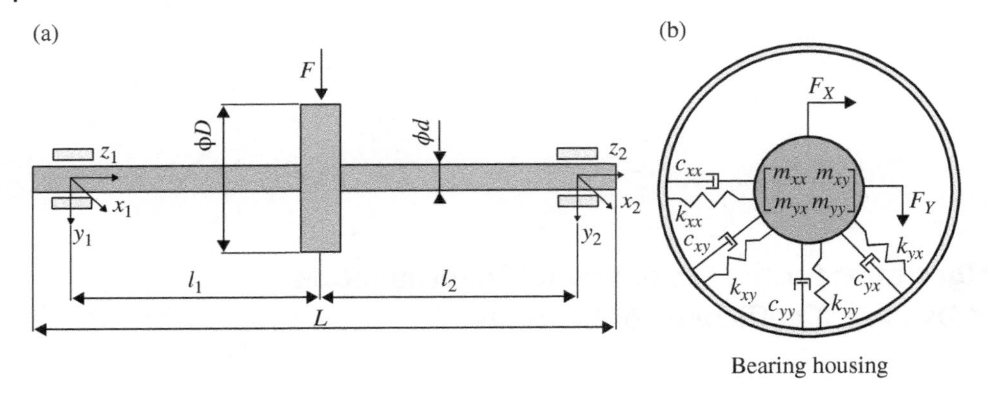

Figure 5.1 (a) Rotor model in the Samcef Rotors software and (b) bearing model.

$$-m_{yy}\omega^2 Y_y - m_{xy}\omega^2 Y_x + k_{yy}Y_y + c_{yy}j\omega Y_y + k_{yx}Y_x + c_{yx}j\omega Y_x = F_y \tag{5.5}$$

$$\begin{bmatrix} k_{xx} - m_{xx}\omega^2 + c_{xx}j\omega & k_{xy} - m_{xy}\omega^2 + c_{xy}j\omega \\ k_{yx} - m_{yx}\omega^2 + c_{yx}j\omega & k_{yy} - m_{yy}\omega^2 + c_{yy}j\omega \end{bmatrix} \cdot \begin{bmatrix} X_x & X_y \\ Y_x & Y_y \end{bmatrix} = \begin{bmatrix} F_x & 0 \\ 0 & F_y \end{bmatrix} \tag{5.6}$$

The first part of Eq. (5.6) is defined as impedance and can be divided into mass, stiffness and damping parameters (5.7):

$$\begin{bmatrix} k_{xx} - m_{xx}\omega^2 + c_{xx}j\omega & k_{xy} - m_{xy}\omega^2 + c_{xy}j\omega \\ k_{yx} - m_{yx}\omega^2 + c_{yx}j\omega & k_{yy} - m_{yy}\omega^2 + c_{yy}j\omega \end{bmatrix} \cdot \begin{bmatrix} 1 & 0 & -\omega^2 & 0 & j\omega & 0 \\ 0 & 1 & 0 & -\omega^2 & 0 & j\omega \end{bmatrix} = \begin{bmatrix} k_{xx} & k_{xy} \\ k_{yx} & k_{yy} \\ m_{xx} & m_{xy} \\ m_{yx} & m_{yy} \\ c_{xx} & c_{xy} \\ c_{yx} & c_{yy} \end{bmatrix} \tag{5.7}$$

The least squares method used to solve the above equation requires a signal in the frequency domain. It is achieved with Fourier transform of the excitation and response signals from the time domain. Vector D is the response signal of the system in the frequency domain, matrix F consisted of elements of excitation force ($f_x(t)$ and $f_y(t)$) in the frequency domain. Taking D and F matrices into account, Eq. (5.6) can be presented as Eq. (5.8). If we consider a dual bearing arrangement (shown in Figure 5.1a), Eq. (5.8) can be represented as Eq. (5.9):

$$\begin{bmatrix} H_{xx_i} & H_{xy_i} \\ H_{yx_i} & H_{yy_i} \end{bmatrix} \cdot \begin{bmatrix} D_{xx_i} & D_{xy_i} \\ D_{yx_i} & D_{yy_i} \end{bmatrix} = \begin{bmatrix} F_{x_i} & 0 \\ 0 & F_{y_i} \end{bmatrix} \tag{5.8}$$

The dynamic stiffness is denoted by vector H. Indices 1 and 2 in Eq. (5.9) denote the first and second bearings, respectively. The index i indicates the successive frequencies analyzed. The first of the double indices indicates the direction of the system response and the second index indicates the direction of the excitation force. For example, element $D^1{}_{yx}$ indicates the vibration of the rotor at the location of the first bearing in the x direction after inducing it by a force acting in the y direction.

$$\begin{bmatrix} H^1_{xx_i} & H^1_{xy_i} \\ H^1_{yx_i} & H^1_{yy_i} \end{bmatrix} = \begin{bmatrix} F^1_{x_i} & 0 \\ 0 & F^1_{y_i} \end{bmatrix} \cdot \begin{bmatrix} D^1_{xx_i} & D^1_{xx_i} \\ D^1_{xx_i} & D^1_{xx_i} \end{bmatrix}^{-1}$$

$$\begin{bmatrix} H^2_{xx_i} & H^2_{xy_i} \\ H^2_{yx_i} & H^1_{yy_i} \end{bmatrix} = \begin{bmatrix} F^2_{x_i} & 0 \\ 0 & F^2_{y_i} \end{bmatrix} \cdot \begin{bmatrix} D^2_{xx_i} & D^2_{xx_i} \\ D^2_{xx_i} & D^2_{xx_i} \end{bmatrix}^{-1} \tag{5.9}$$

Flexibility is defined as the inverse of the dynamic stiffness and can be written using Eq. (5.10):

$$F^1{}_i = \begin{bmatrix} F^1_{xx_i} & F^1_{xy_i} \\ F^1_{yx_i} & F^1_{yy_i} \end{bmatrix} = \begin{bmatrix} H^1_{xx_i} & H^1_{xy_i} \\ H^1_{yx_i} & H^1_{yy_i} \end{bmatrix}^{-1}$$

$$F^2{}_i = \begin{bmatrix} F^2_{xx_i} & F^2_{xy_i} \\ F^2_{yx_i} & F^2_{yy_i} \end{bmatrix} = \begin{bmatrix} H^2_{xx_i} & H^2_{xy_i} \\ H^2_{yx_i} & H^2_{yy_i} \end{bmatrix}^{-1} \tag{5.10}$$

The least squares method can be applied to Eq. (5.10), but it is then necessary to formulate Eq. (5.11). The values of the Z matrix can be determined from Eq. (5.12). In this equation, A_i is a matrix in which the elements are flexibility and frequency signals. It was created by decomposing the real and imaginary parts (5.13). Matrix I is defined by Eq. (5.14), while matrix Z consists of stiffness, damping and mass coefficients of the rotor–bearing system. If we assume that the parameters are determined for a hundred frequency values (this is determined by index i), the matrix A will have the dimensions of 800×12, the matrix I 800×2, and the matrix Z 2×12.

$$A_i \cdot Z = I_i \tag{5.11}$$

$$Z = \left(A_i^T A_i\right)^{-1} A_i^T I_i \tag{5.12}$$

$$A_i = \begin{bmatrix} \text{Real}\left[F_i^1 \cdot \begin{bmatrix} 1 & 0 & 0 & 0 & -(\omega_i)^2 & 0 & 0 & 0 & \omega_i & 0 & 0 & 0 \\ 0 & 1 & 0 & 0 & 0 & -(\omega_i)^2 & 0 & 0 & 0 & \omega_i & 0 & 0 \end{bmatrix} \right] \\ \text{Real}\left[F_i^2 \cdot \begin{bmatrix} 0 & 0 & 1 & 0 & 0 & 0 & -(\omega_i)^2 & 0 & 0 & 0 & \omega_i & 0 \\ 0 & 0 & 0 & 1 & 0 & 0 & 0 & -(\omega_i)^2 & 0 & 0 & 0 & \omega_i \end{bmatrix} \right] \\ \text{Imag}\left[F_i^1 \cdot \begin{bmatrix} 1 & 0 & 0 & 0 & -(\omega_i)^2 & 0 & 0 & 0 & \omega_i & 0 & 0 & 0 \\ 0 & 1 & 0 & 0 & 0 & -(\omega_i)^2 & 0 & 0 & 0 & \omega_i & 0 & 0 \end{bmatrix} \right] \\ \text{Imag}\left[F_i^2 \cdot \begin{bmatrix} 0 & 0 & 1 & 0 & 0 & 0 & -(\omega_i)^2 & 0 & 0 & 0 & \omega_i & 0 \\ 0 & 0 & 0 & 1 & 0 & 0 & 0 & -(\omega_i)^2 & 0 & 0 & 0 & \omega_i \end{bmatrix} \right] \end{bmatrix} \tag{5.13}$$

$$I_i = \begin{bmatrix} 1 & 0 & 1 & 0 & 0 & 0 & 0 & 0 \\ 0 & 1 & 0 & 1 & 0 & 0 & 0 & 0 \end{bmatrix}^T \tag{5.14}$$

The order of damping, mass and stiffness coefficients for the two rotor bearings is given in Eq. (5.15):

$$Z = \begin{bmatrix} k_{xx}^1 & k_{xy}^1 & k_{xy}^2 & k_{xy}^2 & m_{xx}^1 & m_{xy}^1 & m_{xx}^2 & m_{xy}^2 & c_{xx}^1 & c_{xy}^1 & c_{xx}^2 & c_{xy}^2 \\ k_{yx}^1 & k_{yy}^1 & k_{yx}^2 & k_{yy}^2 & m_{yx}^1 & m_{yy}^1 & m_{yx}^2 & m_{yy}^2 & c_{yx}^1 & c_{yy}^1 & c_{yx}^2 & c_{yy}^2 \end{bmatrix}^T \tag{5.15}$$

5.2 Verification of the Calculation Algorithm on the Basis of a Numerical Model

In order to verify the proposed method, a rotor model was built in the Samcef Rotors software (Figure 5.1). The numerical analyses accurately reflected the subsequent experimental research. The model consists of a shaft rotating with a speed of 8000 rpm and bearings modeled by means of stiffness and damping coefficients in the main and cross-coupled directions. During the simulation, the rotor was forced to vibrate with a known value of the excitation force in its central part in the X direction. The displacement of the rotor journals in the X and Y directions was then checked. The simulation was repeated, this time with the same excitation force, but the model was forced in the Y direction. In this case the amplitude of rotor vibrations at the bearing location in the X and Y directions was also measured. When vibrations were measured in the first bearing in the Y direction during the excitation in the X direction, the signal was marked as 1_{XY}. Dynamic coefficients of bearings were calculated on the basis of excitation signals and system responses generated by Samcef Rotors software. The algorithm presented in the previous section was used for calculations. The successive steps of the calculation process are shown in Figure 5.2. Since the stiffness and damping coefficients were previously set in Samcef Rotors, it was possible to compare the values of the coefficients calculated using the proposed algorithm with the actual values.

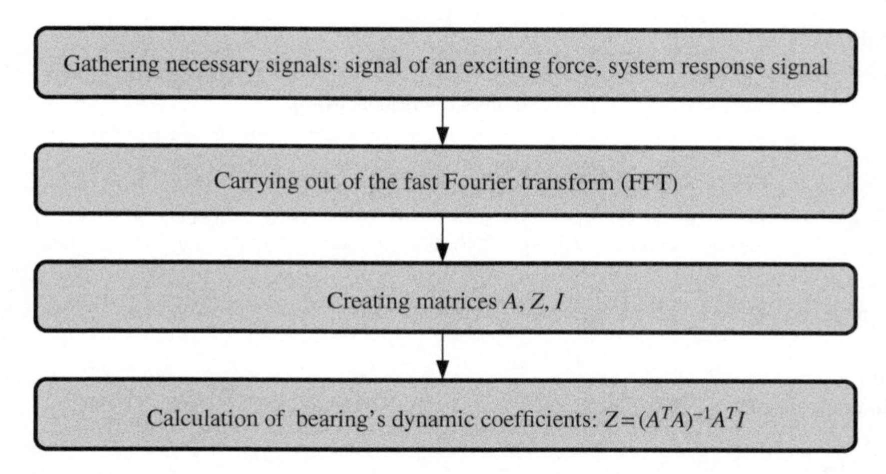

Figure 5.2 Diagram for the calculation of dynamic bearing coefficients.

A symmetrical rotor model was created in the Samcef Rotors software (Figure 5.1a), the parameters of which are listed in Table 5.1. The bearing parameters are the values of stiffness and damping in the main directions xx, yy and cross-coupled directions xy and yx. The set values of stiffness and damping are listed in Tables 5.2 and 5.3. Rotating system modeling, in which bearing parameters are provided by the stiffness and damping coefficients, is a standard way of modeling rotating systems.

In the numerical model, during rotor operation, force F was induced in the X and Y directions. The force signal in the X direction is shown in Figure 5.3a. The same method of excitation is used in experimental research. Then the amplitude of vibrations in the bearings was measured, successively after forcing in the X and Y directions. The vibration amplitude of the first bearing is shown in Figure 5.3b. In the described algorithm for determining stiffness, damping, and mass of the rotor–bearing system the signal in the frequency domain is used, so it is necessary to perform a fast Fourier transform (FFT) of the described signals. System force and response signals for 1200 samples in the second bearing in the frequency domain are shown in Figure 5.4. The function shown in the frequency diagram corresponds to the full range of signal identification (0–87.5 Hz).

Table 5.1 Parameters of the numerical model.

Parameter	Value
Rotor length, L	0.92 m
Distance between bearings, $2 \cdot l_1$	0.58 m
Shaft diameter, d	19.05 mm
Disk diameter, D	152.4 mm
Young's modulus, E	205 109 Pa
Poisson's ratio	0.3
Material density, ρ	7800 kg/m^3

Table 5.2 Stiffness coefficients.

	Stiffness coefficients (N/m)							
	k_{xx}^1	k_{yy}^1	k_{xy}^1	k_{yx}^1	k_{xx}^2	k_{yy}^2	k_{xy}^2	k_{yx}^2
Real values	500 000	450 000	250 000	240 000	550 000	470 000	270 000	260 000
Calculated values Identification range 0–3.125 Hz	499 913	449 922	249 957	239 959	549 905	469 919	269 953	259 955
Relative error % Identification range 0–3.125 Hz	0.02	0.02	0.02	0.02	0.02	0.02	0.02	0.02
Calculated values Identification range 0–87.5 Hz	499 949	449 999	249 630	241 198	550 192	469 822	270 271	259 138
Relative error % Identification range 0–87.5 Hz	0.01	0.00	0.15	0.50	0.03	0.04	0.10	0.33

Table 5.3 Damping coefficients.

	Damping coefficients (N·s/m)							
	c_{xx}^1	c_{yy}^1	c_{xy}^1	c_{yx}^1	c_{xx}^2	c_{yy}^2	c_{xy}^2	c_{yx}^2
Real values	500	550	250	260	550	560	260	270
Calculated values Identification range 0–3.125 Hz	500	550	250	259	550	560	260	271
Relative error % Identification range 0–3.125 Hz	0.00	0.00	0.00	0.38	0.00	0.00	0.00	0.37
Calculated values Identification range 0–87.5 Hz	490	545	248	260	562	565	262	270
Relative error % Identification range 0–87.5 Hz	2.00	0.91	0.80	0.00	2.18	0.89	0.77	0.00

Figure 5.5 shows the results of amplitude and flexibility phase. Flexibility signals are used in this algorithm to calculate the stiffness, damping and mass coefficients of the rotor–bearing system. The resonance of 46 Hz shown in the diagram corresponds to the first form of natural vibrations of the described rotor. The form of the rotor's natural vibrations is shown in Figure 5.6.

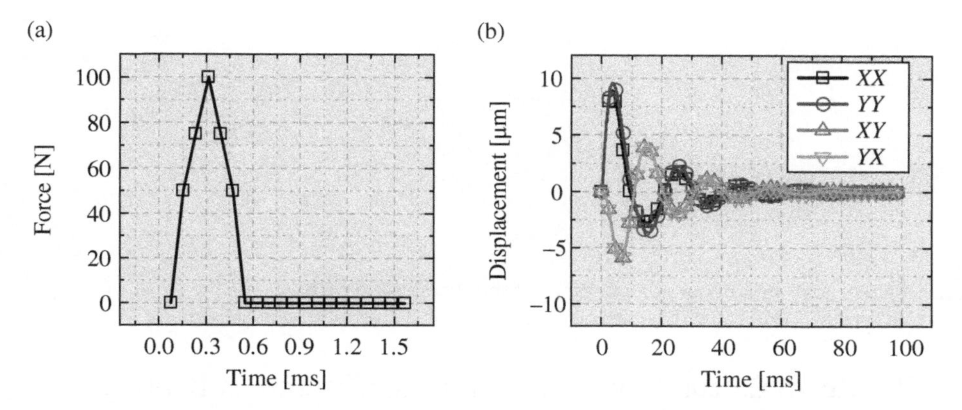

Figure 5.3 (a) Force in direction X as a function of time and (b) displacement of bearing no. 2 as a function of time.

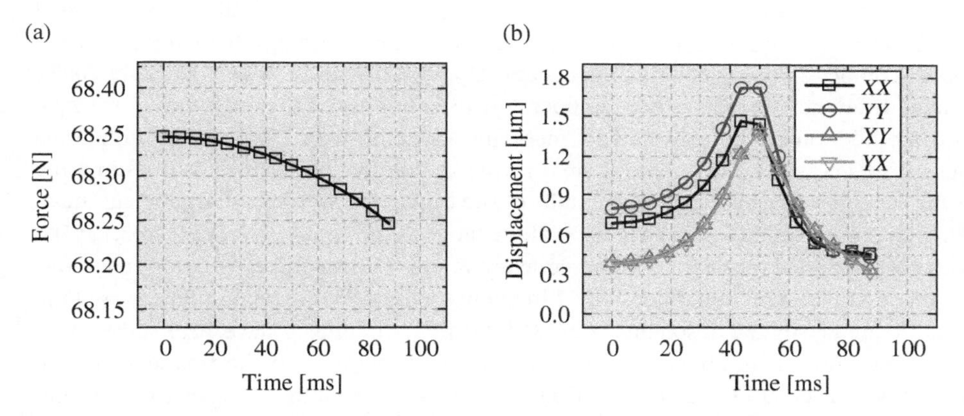

Figure 5.4 (a) Frequency distribution of force and (b) amplitude of node no. 2 as a function of frequency.

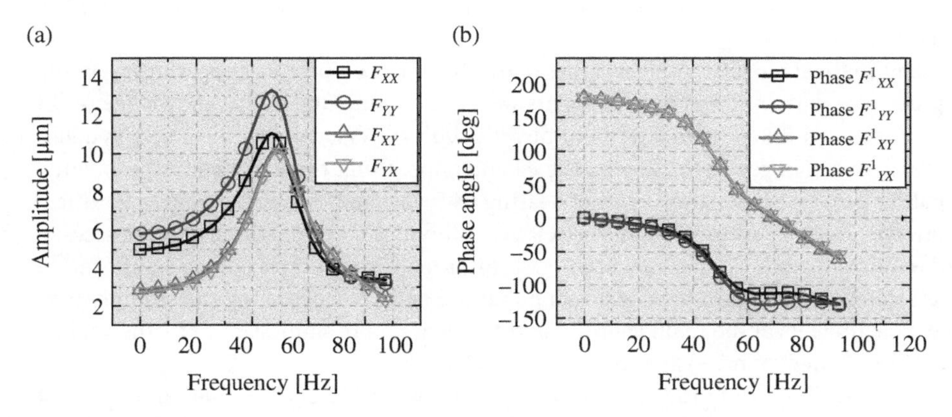

Figure 5.5 Bearing no. 1: (a) flexibility amplitude and (b) flexibility phase.

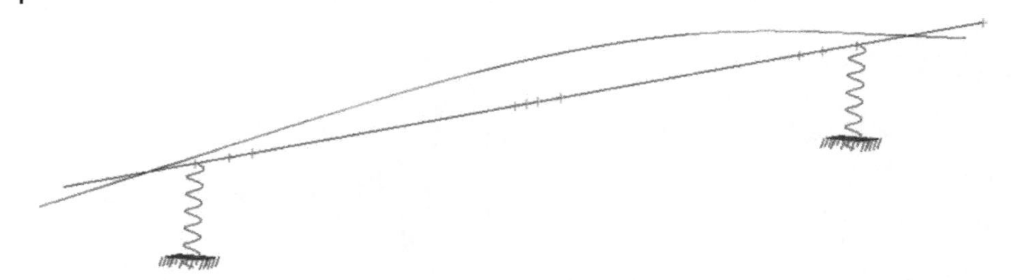

Figure 5.6 The first form of natural vibrations of the rotor at 46 Hz.

5.3 Results of Calculations of Dynamic Coefficients of Bearings

Signals generated by Samcef Rotors software were used to calculate the stiffness, damping and mass coefficients of the rotor–bearing system. A symmetrical rotor model was considered, but different values of stiffness and damping coefficients (including cross-coupled values) were set in the model. In the case of hydrodynamic bearings, the values of cross-coupled contact damping coefficients C_{xy} and C_{yx} have the same values (Kiciński 2005), but here different values have been deliberately used to present the operation of the proposed method. Since the values of stiffness and damping were given in the numerical model, it was possible to compare them directly, and on this basis an error in estimating the coefficients was calculated. The results of the calculated stiffness, damping and mass coefficients are presented for two calculation ranges. The first was from 0 to 3.125 Hz, which corresponded to the first three frequency values. The second calculation range was from 0 to 87.5 Hz. It included the first 57 frequency values. The second range covered the signal frequency in the resonance range. Two different ranges have been taken into account, because in the theoretical case only a small fragment of the frequency of the analyzed signal is sufficient. In experimental practice, researchers suggest taking into account the entire frequency range in which resonances occur (Qiu and Tieu 1997). This procedure ensures higher repeatability of results. The results of calculations of the stiffness, damping and mass coefficients are presented in Table 5.2, Table 5.3, and Table 5.4, respectively.

In the numerical model, the exact values of the dynamic coefficients of bearings were set. The process of determining the coefficients was carried out. Due to the knowledge of set values, it was possible to compare them directly with the calculated values and to determine relative errors generated during the calculations, using the described experimental method. The relative estimation error of bearing stiffness coefficients in the first identification interval does not exceed 0.02%. The relative estimation error of stiffness in the second, larger frequency range does not exceed 0.5%. This is a fairly accurate estimation of bearing stiffness coefficients, especially considering that the error in calculating the second range stiffness coefficients at the main coordinates (which are more significant in further modeling of the system dynamics) does not exceed 0.04%.

The error resulting from the calculation of damping coefficients is small and does not exceed 0.4% in the first calculation range. This error occurs only during the cross-coupled

Table 5.4 Mass coefficients.

	Mass coefficients (kg)								
	m^1_{xx}	m^1_{yy}	m^1_{xy}	m^1_{yx}	m^2_{xx}	m^2_{yy}	m^2_{xy}	m^2_{yx}	Σm
Real values									4.845
Calculated values Identification range 0–3.125 Hz	2.35	2.40	−0.02	0.01	2.51	2.45	0.02	−0.01	4.851
Relative error % Identification range 0–3.125 Hz									0.12
Calculated values Identification range 0–87.5 Hz	2.33	2.40	−0.02	0.04	2.54	2.45	0.02	−0.03	4.855
Relative error % Identification range 0–87.5 Hz									0.21

calculation of the damping coefficients. The error in calculating the damping coefficients for the second calculation range reaches a maximum of approximately 2%.

The mass of the shaft modeled in the Samcef Rotors software is 4.845 kg. The mass of the shaft calculated on the basis of the algorithm presented is 4.85 kg for the first identification range and 4.86 kg for the large identification range. The mass is calculated by adding the mass coefficients on the main diagonals (*xx* and *yy*) and dividing this value by two. The interpretation of the values of the mass coefficients is as follows: the $m^1{}_{xx}$ coefficient is the mass of the shaft parts involved in vibrations in the *X* direction per first bearing after forcing the system in the *X* direction. In contrast, $m^1{}_{yy}$ is the mass of the shaft parts involved in the *Y* direction vibrations per first bearing after forcing the system in the *Y* direction. Dividing the sum of the mass coefficients by two to calculate the mass of the shaft is used because the system was induced twice. The cross-coupled mass coefficients (*xy*, *yx*) should equal zero. Errors in the calculation of mass coefficients for the first and second identification ranges are 0.12 and 0.21%, respectively. These results mean that the rotor mass of 4.845 kg can be determined with an accuracy of 0.01 kg (for the second calculation range, where the greater relative error is obtained).

During the verification of the proposed algorithm, the mass, stiffness and damping coefficients of the systems of a rotor and two bearings were calculated for the two identification ranges. The first range of identification covered only the first few frequencies, while the second covered the entire resonance range. When extending the calculation range, the error usually increases slightly, but still remains at a relatively low level. Extension of the identification range ensures less influence of additional factors on the values of calculated mass, stiffness and damping coefficients during experimental research. The author recommends that in the case of repeatable results, a narrow range of identification should be used. In experimental research, a better repeatability of results is ensured by analyses carried out on the basis of a signal with a wider frequency range.

5.4 Summary

This chapter describes how to extend the method of determining 16 stiffness and damping coefficients of hydrodynamic bearings, proposed by Qiu and Tieu (1997), by determining a further eight mass coefficients. All coefficients are calculated in one operation using the least squares method. The calculated mass coefficients correspond to the mass of the rotor involved in bearing vibrations.

In order to verify the described method, a numerical model was built in the Samcef Rotors software. The rotor was modeled with bearings set as stiffness and damping coefficients. Then the process of identification of these coefficients was carried out. The values calculated in this way were compared with the set values as bearing definitions.

Calculations of dynamic coefficients of bearings were made for two identification ranges. The first covered only the initial frequency range and the second covered the entire frequency range in which the rotor resonance occurred. It turns out that an increase in the identification range usually results in a slight increase in the error rate, but still remains at an acceptable level. Extending the identification range to the resonance range may increase the repeatability of results in experimental research. The absolute error of the calculated coefficients in most cases did not exceed 1%, which was considered to be very high compliance.

6

Inclusion of the Impact of an Unbalanced Rotor

Chapter 5 presented the calculations for the rotating shaft, but without residual unbalance. In practice, however, there is a residual unbalance in the rotor of every rotating machine because it is impossible to balance the rotor perfectly. The residual unbalance must be included in the calculation. This chapter presents the influence of the rotating shaft with residual unbalance on the determination of the stiffness, damping and mass coefficients. As in the previous chapter, the calculations were made on the basis of the signal generated by the numerical model built in the Samcef Rotors software. The chapter also presents the way the error changes in the calculation of the stiffness, damping and mass coefficients, both with the unbalance and when its effect is leveled out. Some of the results in this chapter are also presented elsewhere (Breńkacz 2015b).

6.1 Calculation Scheme

The calculations presented in this chapter are prepared for a system consisting of a rotor and two bearings (Figure 5.1a). During experimental studies, on the basis of which the stiffness, damping and mass coefficients of the rotor–bearing system were determined, it is necessary to use a impact hammer twice for excitation. In the first step it is necessary to induce the rotor in the X direction in its central part using the F_X force, after which the vibrations of the system need to stabilize. In the second step, the system must be induced in the Y direction (perpendicular to the X direction) by the F_Y excitation force. The system needs to return to its stable operation after this step as well. Knowing the value of the force signal and the response of the rotating system at the place where the bearings are located, it is possible to calculate its damping, stiffness and mass coefficients.

The response of the system after excitation (displacement signal in bearings) consists of two parts. The first part is related to the reaction to excitation. The second part is related to the existence of constant rotational speed and residual unbalance. In order to correctly calculate the dynamic coefficients of bearings, it is necessary to subtract from the response signal, the second part – related to residual unbalance. In order to do so, the calculation scheme shown in Figure 5.2 needs to be modified by a related operation (Figure 6.1).

Bearing Dynamic Coefficients in Rotordynamics: Computation Methods and Practical Applications, First Edition. Łukasz Breńkacz.

Companion website: www.wiley.com/go/brenkacz/bearingdynamiccoefficients

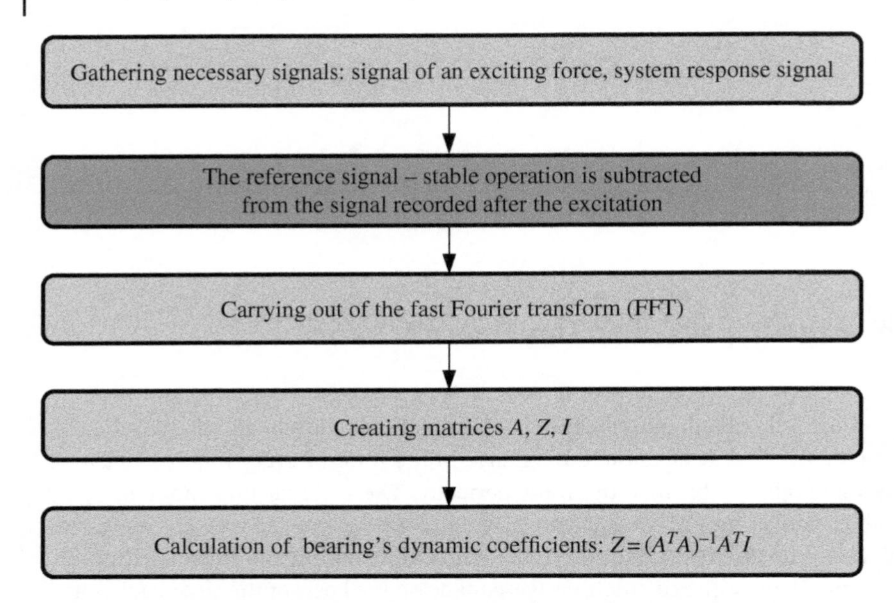

Figure 6.1 Calculation scheme of dynamic coefficients of bearings including unbalance.

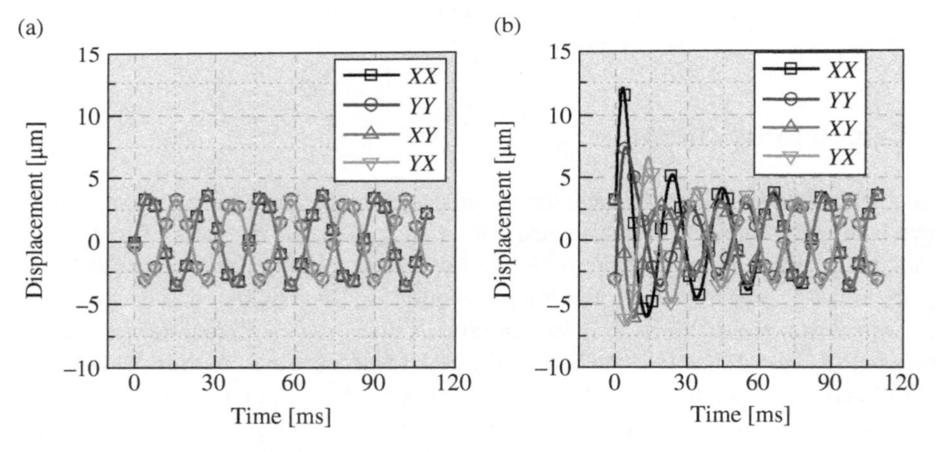

Figure 6.2 (a) Stable operation of the rotor – reference signal in the bearing and (b) amplitude of vibrations after inducing the rotor by means of a impact hammer – signal in the bearing.

The unbalance adopted in the numerical model is 1 g within a radius of 10 mm. This unbalance was adopted in accordance with PN-93/N-01359 (1993) for a rotor with a mass of 4.844 kg, balanced in balance quality grade G 6.3 for a rotational speed of 10 000 rpm.

The stable operation of the rotor rotating at 2800 rpm is shown in Figure 6.2a. It is a signal generated during stable operation and measured at the bearing base – reference signal. When the unbalance increases, the amplitude of the signal increases as well. When the rotational speed is increased, vibrations shall be recorded at a higher frequency. Figure 6.2b shows the response signal of the system after the application of the excitation force. This

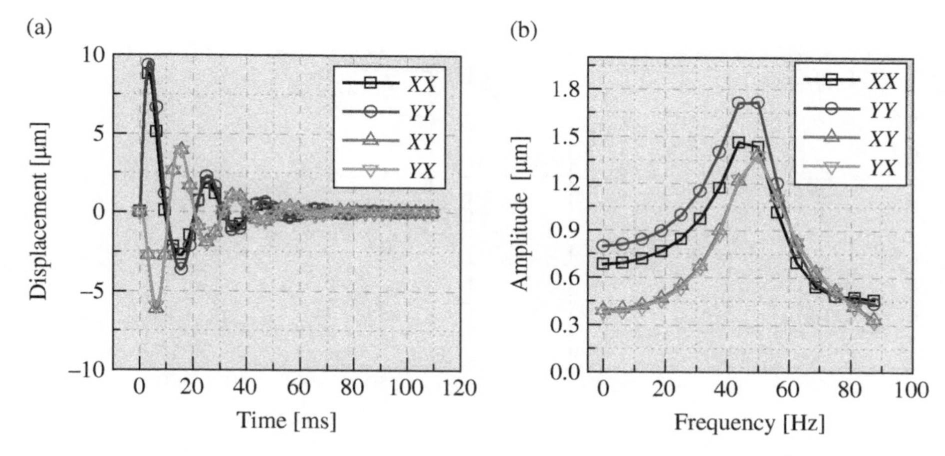

Figure 6.3 (a) Amplitude of vibrations after subtracting the reference signal from the excitation signal and (b) fast Fourier transform of this signal.

figure shows an increase in vibrations caused by the excitation force and stabilization of the system operation. The signal, after subtracting the reference signal from the signal recorded after excitation, is shown in Figure 6.3. Further calculations of stiffness, damping and mass coefficients are carried out using this signal as the response of the system. This signal (1200 samples), after the application of fast Fourier transform (FFT), is shown in Figure 6.3b.

The excitation force used to induce the system in its central part is shown in Figure 5.3a. The same force F_X, $F_Y = [0, 50, 75, 100, 75, 50, 0]$ N was used in two perpendicular directions. In the Samcef Rotors software, an analysis was made in the time domain, where the time increment is 1/12 800 seconds. The curve of the excitation force in the frequency domain is shown in Figure 5.4a.

6.2 Definition of the Scope of Identification

To determine the mass, stiffness and damping coefficients of the rotor–bearing system, it is necessary to select an appropriate identification range. This is the frequency range used in the algorithm for calculations. Ideally, the first few frequencies are sufficient enough, but in experimental practice this range is extended to the entire resonance range. This ensures higher repeatability of experimental results (Nordmann and Schoellhorn 1980). If the rotational speed related component is outside the resonant speed range, a shorter interval may be used as the identification range. Figure 6.4 shows the FFT diagram of the system response at the location of bearings for two rotor speeds (2800 and 10 000 rpm). Analyzing the two diagrams, you can see that in the first case the rotational speed related component is within the identification range, and in the second case it is outside the identification range. In the first case, without the removal of the speed related component, the calculation results are subject to an error of approximately 300%, and thus are unusable in practice. In the second case, if the identification range does not include the speed (it is outside the identification range), this error reaches a maximum of approximately 6%, but usually equals approximately 3%.

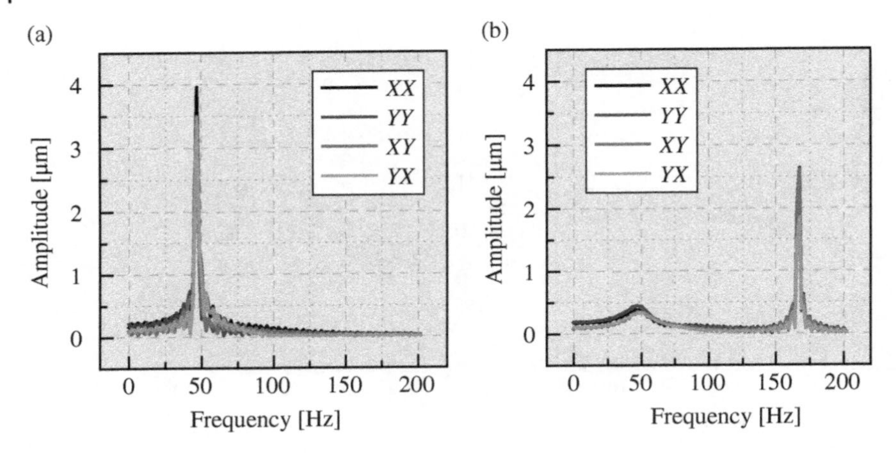

Figure 6.4 The fast Fourier transform displacement signal in bearings for rotational speed (a) 2800 rpm and (b) 10 000 rpm.

6.3 Results of the Calculation of Dynamic Coefficients of Bearings Including Rotor Unbalance

In the analyzed case, 1000 samples from a time signal were used. Assuming that the bearing parameters are determined for eight frequency values (as indicated by index i), the dimensions of the A matrix were 64×12, the I matrix 64×2, and the Z matrix 2×12.

In order to calculate the stiffness, damping and mass coefficients of the rotor–bearing system, it is necessary to determine the flexibility. It is obtained by multiplying the signal of the excitation force and the system response in the frequency domain. The diagram of amplitude and phase angle of flexibility is shown in Figure 6.5. The phase angle is an argument for the sinusoidal function and determines in which part of the wave period the point of the sinusoidal function is located. Calculations were carried out for a rotational speed of

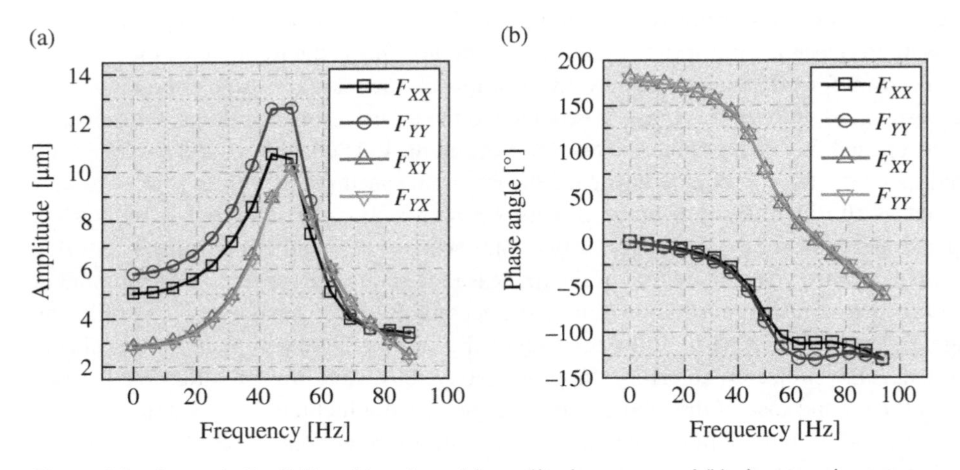

Figure 6.5 Dynamic flexibility of bearings: (a) amplitude course and (b) phase angle course.

2800 rpm, after removing the constant component of the signal related to the unbalance (based on the data from Figure 6.3 and the previously described forced force signal). For calculations of stiffness and mass coefficients carried out on the basis of such prepared data, the error does not exceed 1%. It is slightly higher for damping coefficients and reaches a maximum of 2%.

6.4 Summary

This chapter presents how the measurement error related to the determination of dynamic coefficients of bearings including the unbalance changes. In order to obtain correct results, it is necessary to subtract a constant component associated related to residual unbalance from the system's response signal. If in the frequency domain the signal component related to the rotational speed is outside the identification range (resonance range), it is possible to include a shorter signal identification interval – without frequencies associated with rotational speed.

On the basis of the model under consideration, it can be stated that if the curve coming from the rotational speed is not subtracted from the tested signal and it is outside the identification range, there will be an error of calculation of dynamic coefficients of bearings, usually not exceeding 3%. A good way to eliminate the effect of residual unbalance from the results is to subtract a constant component (understood in this case as the displacement of the rotor journals during its stable operation, which is mainly caused by the unbalance of the rotor) from the signal after inducing. Such a procedure carried out on the displacement signal after impulse excitation means that the error in calculating the stiffness and mass coefficients of the rotor–bearing system usually does not exceed 1%, and the damping coefficients do not exceed 2%. If the speed related component of the FFT spectrum coincides with the resonance frequencies on the basis of which it is intended to calculate the stiffness and damping coefficients, a calculation error of up to several hundred percent will be obtained. The smallest calculation error is obtained by subtracting the reference signal from the signal recorded after the excitation. This results in a calculation accuracy similar to that of the signal without the residual unbalance. During the experimental research, the subtraction operation was performed for each signal separately. As it is quite time-consuming, the "Signal" software was developed to aid the operations. The software was written in the Matlab environment and is described in Section 8.1.

7

Sensitivity Analysis of the Experimental Method of Determining Dynamic Coefficients of Bearings

The impulse method described in the two previous chapters was used to evaluate the sensitivity of the method for determining the mass, damping and stiffness coefficients of the rotor–bearing system. The literature describes the influences of many parameters on the results of calculations of experimental dynamic coefficients of bearings, such as supply pressure or bearing geometry. However, there is no description of the sensitivity of the experimental method itself.

Sensitivity analysis is an analysis based on the prediction of results, using variable systems that affect the results. It is an important tool for risk reduction and can be used to estimate the potential impact of different parameters on the final result of the calculations. The construction of a numerical model of the rotor in the Samcef Rotors software made it possible to check the influence of input parameters, such as the variable position and angle of the excitation force and the displacement of sensors measuring the displacement of bearing journals. The influence of changes in rotor material stiffness, unbalance, and geometry on the values of calculated stiffness, damping and mass coefficients of the rotor–bearing system was also checked.

It turns out that the stiffness, damping and mass coefficients are determined with different accuracy. Different results were also obtained for the main and cross-coupled coefficients. Signals of excitation force and response of the system generated during changes in these parameters are presented. This chapter also includes information on improving the accuracy of the calculations and the results obtained after appropriate revisions. Some of the results in this chapter are also presented elsewhere (Breńkacz and Żywica 2016a).

7.1 Method of Carrying Out a Sensitivity Analysis

The sensitivity analysis was not aimed at checking all possible parameters in a wide range of cases, but to identify those that have a significant impact on the results of the calculations. Figure 7.1 shows a diagram according to which the sensitivity of the method was checked. The left side of the figure shows the parameters that change during the determination of the dynamic coefficients of bearings. The middle part shows the calculation scheme used in this method. The right side of the figure shows that the results of the

Bearing Dynamic Coefficients in Rotordynamics: Computation Methods and Practical Applications,
First Edition. Łukasz Breńkacz.

Companion website: www.wiley.com/go/brenkacz/bearingdynamiccoefficients

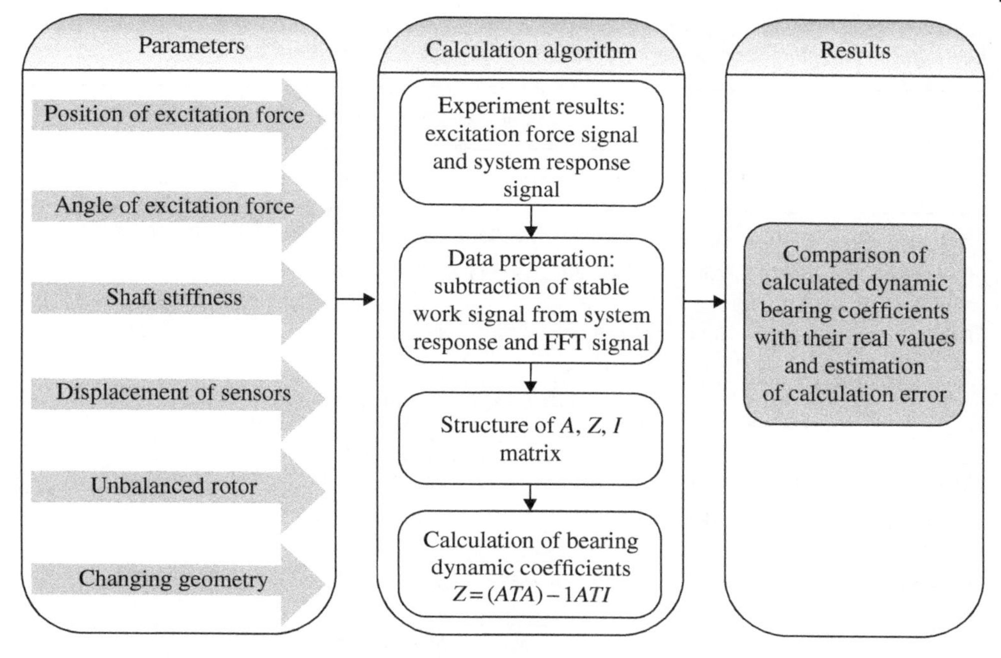

Figure 7.1 Diagram showing the sensitivity analysis scheme.

dynamic coefficients of bearings for different input parameters have been compiled and a calculation error has been estimated. As in the numerical model of the rotor, the bearing stiffness and damping coefficients are known; it is possible to directly compare the calculated coefficients with the real values.

7.2 Description of the Reference Model

The results of calculations together with the initial relative error are presented in Table 7.1. They differ slightly from those presented in Tables 5.2–5.4, because the operations were performed on the signal after removing the constant component. The rotor speed was 2800 rpm. Therefore it is a case where the speed corresponds to the resonance speed, and the component of the signal associated with the speed is in the identification range (resonance range) under consideration. Thus, it is the most unfavorable case from the point of view of accuracy of calculations.

7.3 Influence of the Stiffness of the Rotor Material

One of the analyses was to change the stiffness of the rotor material. The value of Young's modulus was increased one hundred times. The purpose of this procedure is to check whether the material stiffness affects the calculation results. It turned out that after this test, the same calculation results were obtained in two cases. This means that for materials

Table 7.1 Summary of the real and calculated stiffness, damping and mass coefficients for the two bearings for the reference case; representation of the relative calculations error.

	Stiffness coefficients (N/m)								
	k^1_{xx}	k^1_{yy}	k^1_{xy}	k^1_{yx}	k^2_{xx}	k^2_{yy}	k^2_{xy}	k^2_{yx}	
Real values	500 000	450 000	250 000	240 000	550 000	470 000	270 000	260 000	
Calculated values	498 232	450 488	248 338	240 544	548 107	470 504	268 288	260 583	
Relative error %	0.35	0.11	0.66	0.23	0.34	0.11	0.63	0.22	
	Damping coefficients (N·s/m)								
	c^1_{xx}	c^1_{yy}	c^1_{xy}	c^1_{yx}	c^2_{xx}	c^2_{yy}	c^2_{xy}	c^2_{yx}	
Real values	500	550	250	260	550	560	260	270	
Calculated values	507.5	547.7	259.4	258.8	558.1	558.3	269.2	270.7	
Relative error %	1.50	0.42	3.76	0.46	1.47	0.30	3.54	0.26	
	Mass coefficients (kg)								
	m^1_{xx}	m^1_{yy}	m^1_{xy}	m^1_{yx}	m^2_{xx}	m^2_{yy}	m^2_{xy}	m^2_{yx}	Mass
Real values	2.423	2.423	0.000	0.000	2.423	2.423	0.000	0.000	4.845
Calculated values	2.338	2.404	−0.071	−0.029	2.497	2.451	−0.035	−0.053	4.845
Relative error %	3.49	0.77	—	—	3.07	1.18	—	—	0.003

with a Young's modulus (210 GPa) as for steel or higher – and these shaft materials are the most common – the rigidity does not affect the calculation results for the bearing dynamic coefficients.

7.4 Influence of Uneven Force Distribution on Two Bearings

The described algorithm in the basic version assumes a symmetrical division of the excitation force onto two bearings. This assumption requires inducing the rotor in its central part, but it should be borne in mind that it is not possible to do this in every rotating machine. It turned out that in the described model the displacement of the excitation force, even as small as by 5% in relation to the distance between the bearings, can lead to the calculation of dynamic coefficients of bearings with an error of several percent. Appropriate consideration of the proportional distribution of excitation force between two bearings results in correct calculation results.

Figure 7.2a shows a diagram with the displacement of the excitation force marked. This displacement in the described case was $s = 30$ mm. This value is 5% of the shaft's length between bearings. The results were compared with the force F_{ref} located in the middle section of the rotor. Figure 7.2b shows the response signal of the system in the *X* direction after inducing it in the *X* direction by the force F_{ref} (XX_{ref}) and after inducing it with the force *F* shifted by the s distance from the center of the system (*XX*). The signal is only shown in the first direction, as it is the one with the most noticeable difference.

In the algorithm of calculation of dynamic coefficients of bearings, a correction can be applied to improve the accuracy of calculations. The values of forces per bearing shall be calculated from geometric proportions using the dimensions l_{1f} and l_{2f} described by Eqs. (7.1) and (7.2):

$$l_{1f} = l_1 + s = 290 + 30 = 320 \tag{7.1}$$

$$l_{2f} = l_2 - s = 290 - 30 = 260 \tag{7.2}$$

Force correction consists of multiplying each element of the force vector by the values calculated from relationship (7.3). Values of the force divided proportionally onto two bearings should be entered into Eq. (5.9):

$$F_{1\mathrm{shifted}} = F_1 \cdot \frac{l_{2f}}{2\,l_1}, \quad F_{2\mathrm{shifted}} = F_2 \cdot \frac{l_{1f}}{2\,l_2} \tag{7.3}$$

The summary of stiffness, damping and mass coefficients calculated for the case with the displaced force is shown in Table 7.2. If the calculation does not include a correction related to the displacement of the excitation force, we obtain the results of the calculation of stiffness, damping and mass coefficients burdened with a relative error of up to 12% for stiffness, 15.5% for damping and 1.73% for mass. After the correction of the displacement of the excitation force, values of dynamic coefficients are calculated with an error in determining stiffness coefficients of less than 0.73%, damping coefficients of less than 3.73% and mass

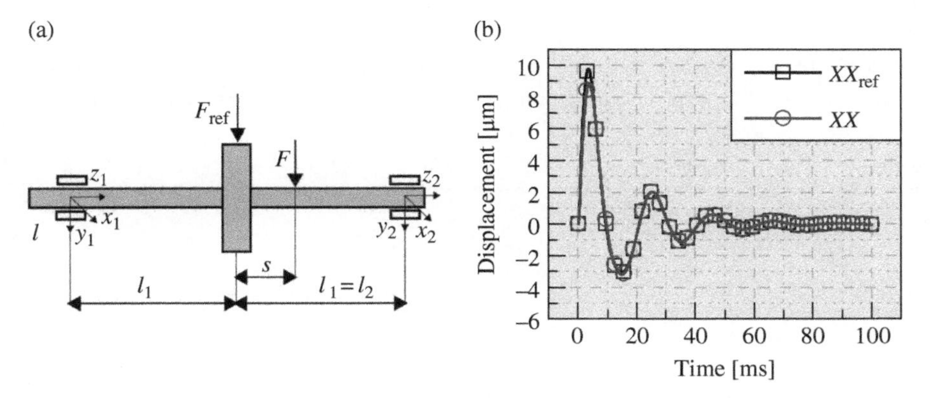

Figure 7.2 (a) Rotor model on which the method of force displacement is marked. (b) Bearing response signal for the first bearing in the *X* direction after inducing the system in the *X* direction; comparison after excitation in the middle and force displaced by the value $s = 30$ mm.

Table 7.2 Summary of the real and calculated stiffness, damping, and mass coefficients for the two bearings for a case with displaced force; results "with correction" take into account the uneven distribution of excitation force in the calculation procedure.

	Stiffness coefficients (N/m)								
	k^1_{xx}	k^1_{yy}	k^1_{xy}	k^1_{yx}	k^2_{xx}	k^2_{yy}	k^2_{xy}	k^2_{yx}	
Real values	500 000	450 000	250 000	240 000	550 000	470 000	270 000	260 000	
Calculations without correction	555 537	502 566	276 803	268 393	496 849	426 325	243 265	236 089	
Relative error %	11.11	11.68	10.72	11.83	9.66	9.29	9.90	9.20	
Calculations with correction	498 068	450 577	248 168	240 628	548 248	470 428	268 430	260 512	
Relative error %	0.39	0.13	0.73	0.26	0.32	0.09	0.58	0.20	
	Damping coefficients (N·s/m)								
	c^1_{xx}	c^1_{yy}	c^1_{xy}	c^1_{yx}	c^2_{xx}	c^2_{yy}	c^2_{xy}	c^2_{yx}	
Real values	500	550	250	260	550	560	260	270	
Calculations without correction	567.0	610.6	288.7	290.2	505.1	506.1	244.4	243.9	
Relative error %	13.40	11.02	15.48	11.62	8.16	9.63	6.00	9.67	
Calculations with correction	508.3	547.4	258.9	260.2	557.4	558.5	269.7	269.2	
Relative error %	1.66	0.47	3.56	0.08	1.35	0.27	3.73	0.30	
	Mass coefficients (kg)								
	m^1_{xx}	m^1_{yy}	m^1_{xy}	m^1_{yx}	m^2_{xx}	m^2_{yy}	m^2_{xy}	m^2_{yx}	*Mass*
Real values	2.423	2.423	0.000	0.000	2.423	2.423	0.000	0.000	4.845
Calculations without correction	2.778	2.868	−0.089	−0.031	2.123	2.089	−0.031	−0.045	4.929
Relative error %	14.68	18.38	—	—	12.38	13.75	—	—	1.733
Calculations with correction	2.491	2.571	−0.080	−0.028	2.342	2.306	−0.035	−0.049	4.855
Relative error %	2.82	6.13	—	—	3.31	4.83	—	—	0.204

coefficients of approximately 0.2%. It should be mentioned at this point that the accuracy of the determination of stiffness and damping coefficients in the main directions is higher than the accuracy of the determination of coefficients in cross-coupled directions.

It is worth noting that most often it is the mass of the shaft that is the parameter known before starting the calculation, while the stiffness and damping coefficients are unknown. After performing the calculations and receiving the set of dynamic coefficients of bearings, it is possible to evaluate the error of calculating the mass coefficients, which at the first stage of calculation can be used to verify the results obtained. In the model under consideration, the expected value of mass coefficients in the main directions *xx* and *yy* is half the weight of the shaft. Comparing this value with the calculated value it can be stated that when the calculations for force are performed without correction (i.e. not including unequal distribution of force in the algorithm of calculations), the relative error is in the range of 12.4–18.4%. When force correction is taken into account, errors in the calculation of individual mass factors range from 2.8 to 4.8%.

7.5 Changing the Direction of the Excitation Force and its Effect on the Results Obtained

Experimental tests are carried out by inducing the rotating shaft twice with the use of a impact hammer in directions perpendicular to the axis of the rotor. In the previous subsection an example was described, from which it can be concluded that in order to obtain correct results it is necessary to adopt a proper definition of excitation force. In this subsection a case will be described of when a mistake is made in experimental research consisting in hitting a rotating rotor at a certain angle in relation to the intended direction. In order to check this effect, the numerical model uses a force $F_{x\alpha}$ in an axis offset at an angle of 15° from the expected direction. The F_y impact was set along the Y axis. Figure 7.3a shows a diagram of the action of the rotated excitation force. Figure 7.3b shows the result of the system response in the first bearing. Only the response of the system in the X and Y directions is presented after excitation in the X direction. The biggest difference in results is

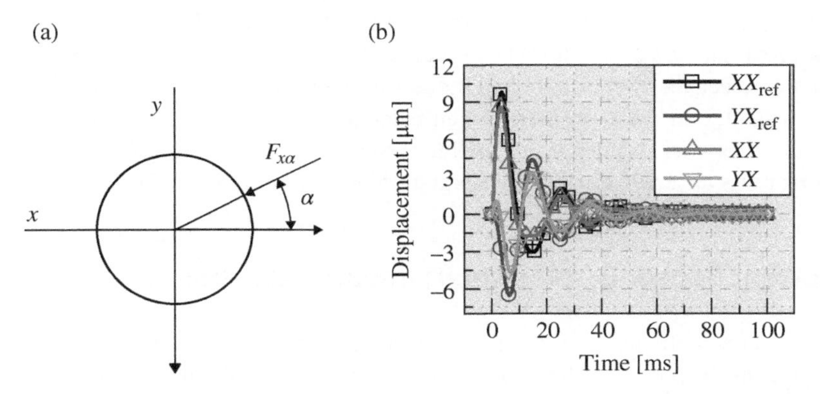

Figure 7.3 (a) Model showing how the force is applied at an angle and (b) the response signal of the first bearing in the X and Y directions after excitation at an angle of $\alpha = 0°$ and $\alpha = 15°$.

Table 7.3 Summary of the real and calculated stiffness, damping and mass coefficients for the two bearings for a case with an excitation at an angle of $\alpha = 15°$.

	Stiffness coefficients (N/m)							
	k^1_{xx}	k^1_{yy}	k^1_{xy}	k^1_{yx}	k^2_{xx}	k^2_{yy}	k^2_{xy}	k^2_{yx}
Real values	500 000	450 000	250 000	240 000	550 000	470 000	270 000	260 000
Calculated values	516 049	383 855	257 221	106 858	567 699	398 520	277 880	113 519
Relative error %	3.21	14.70	2.89	55.48	3.22	15.21	2.92	56.34
	Damping coefficients (N·s/m)							
	c^1_{xx}	c^1_{yy}	c^1_{xy}	c^1_{yx}	c^2_{xx}	c^2_{yy}	c^2_{xy}	c^2_{yx}
Real values	500	550	250	260	550	560	260	270
Calculated values	523.9	478.5	267.8	123.4	576.2	486.5	277.9	121.7
Relative error %	4.78	13.00	7.12	52.54	4.76	13.13	6.88	54.93
	Mass coefficients (kg)							
	m^1_{xx}	m^1_{yy}	m^1_{xy}	m^1_{yx}	m^2_{xx}	m^2_{yy}	m^2_{xy}	m^2_{yx}
Real values	2.423	2.423	0.000	0.000	2.423	2.423	0.000	0.000
Calculated values	2.426	2.424	−0.071	−0.065	2.591	2.462	−0.034	−0.072
Relative error %	0.16	0.07	—	—	6.95	1.62	—	—

visible on these courses. XX_{ref} and YX_{ref} indicate the response of the system when the force F_x acts in the X axis. Signals XX and YX are generated by force $F_{x\alpha}$ acting at an angle of 15°. It should also be emphasized that the accepted angle of excitation is quite large and should be minimized during experimental studies.

Table 7.3 presents the results for stiffness, damping and mass coefficients calculated from the signal generated after inducing the system in the X direction by a force rotated at an angle of $\alpha = 15°$. In the calculation algorithm, the whole force value was treated as a force set at an angle of $\alpha = 0°$. Errors in the calculation of stiffness coefficients can exceed 56%, of damping coefficients approximately 55%, whereas mass of the shaft is determined with an accuracy of approximately 2.2%. It should be noted that the error in determining some coefficients does not exceed 3%, therefore not all the results are burdened with such a large error. The largest calculation error occurred when calculating YX cross-coupled coefficients. A significant error was also generated during the calculation of YY coefficients. A relatively small error was observed during the calculation of XX and XY coefficients.

7.6 Eddy Current Sensor Displacement Impact Assessment

Calculations of dynamic coefficients of bearings are made on the basis of the signal of excitation force and the response of the system within bearings. Since in experimental research it is usually impossible to place sensors in bearing supports, it is necessary to check the influence of placing eddy current sensors next to bearing supports on the results obtained. A diagram of the arrangement of bearing supports and measuring sensors is

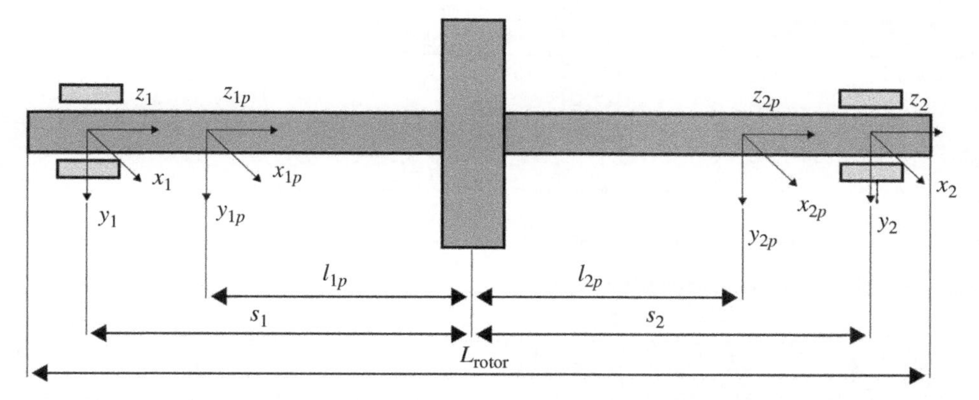

Figure 7.4 Model on which the place of bearing displacement measurement was marked (systems with *p* indexes) and places taken into account during calculation of dynamic coefficients of bearings (means of bearing housings).

shown in Figure 7.4. Dimensions s_1 and s_2 determine the distance of bearing supports from the center of the system. Distances l_{1p} and l_{2p} determine the distance of measurement sensors from the center of the system. The length of the rotor is shown as L_{rotor}.

It turns out that if calculations are carried out for measuring points displaced by only 30 mm (i.e. 5% of the distance between supports), the results of stiffness, damping and mass coefficients are calculated with a relative error of approximately 2%. In order to improve the accuracy of calculations, Eqs. (7.4) and (7.5) can be used. After multiplying the system response signal, which was measured at the sensor mounting point, by the following equations we obtain accurate values.

$$\begin{bmatrix} XX_1 & YY_1 \\ XX_1 & YY_2 \end{bmatrix} = \frac{1}{s_1 + l_2} \cdot \begin{bmatrix} s_2 + l_1 & s_1 - l_1 \\ s_2 - l_2 & s_1 + l_2 \end{bmatrix} \cdot \begin{bmatrix} XX_{1p} & YY_{1p} \\ XX_{2p} & YY_{2p} \end{bmatrix} \tag{7.4}$$

$$\begin{bmatrix} XY_1 & YX_1 \\ XY_1 & YX_2 \end{bmatrix} = \frac{1}{s_1 + l_2} \cdot \begin{bmatrix} s_2 + l_1 & s_1 - l_1 \\ s_2 - l_2 & s_1 + l_2 \end{bmatrix} \cdot \begin{bmatrix} XY_{1p} & YX_{1p} \\ XY_{2p} & YX_{2p} \end{bmatrix} \tag{7.5}$$

7.7 Calculation Results for an Asymmetrical Rotor

It is often necessary to determine the dynamic coefficients of bearings supporting an asymmetrical rotor. This chapter presents the calculations for the example of an asymmetrical rotor. For this purpose, the length of one of the rotor ends was shortened by 60 mm. Such cases occur when, for example, one of the shaft ends is longer to be able to mount a coupling on it. A schematic view of the rotor is shown in Figure 7.5. The dimensions of the rotor for this calculation are: $l_1 = l_2 = 290$ mm, $L_1 = 460$ mm, and $L_2 = 400$ mm. A diagram of the response of the system in the first bearing after inducing it between the bearings is shown in Figure 7.6. This change in geometry results means the mass of the rotor is 4.71 kg.

Table 7.4 lists mass, damping and stiffness coefficients for an asymmetrical rotor. The error in calculating stiffness coefficients does not exceed 0.7% and for damping coefficients

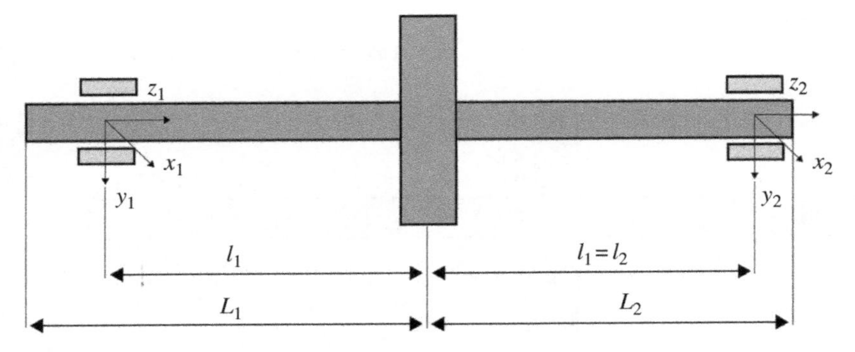

Figure 7.5 Diagram of an asymmetrical rotor.

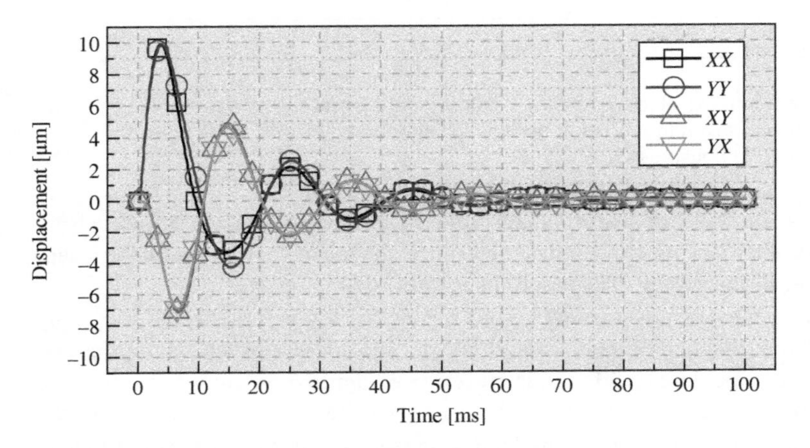

Figure 7.6 Response of an asymmetrical rotor system.

Table 7.4 Summary of the real and calculated stiffness, damping and mass coefficients of two bearings for an asymmetrical rotor.

	Stiffness coefficients (N/m)							
	k^1_{xx}	k^1yy	k^1_{xy}	k^1_{yx}	k^2_{xx}	k^2_{yy}	k^2_{xy}	k^2_{yx}
Real values	500 000	450 000	250 000	240 000	550 000	470 000	270 000	260 000
Calculated values	498 198	450 495	248 331	240 585	548 158	470 442	268 370	260 557
Relative error %	0.36	0.11	0.67	0.24	0.33	0.09	0.60	0.21
	Damping coefficients (N·s/m)							
	c^1_{xx}	c^1_{yy}	c^1_{xy}	c^1_{yx}	c^2_{xx}	c^2_{yy}	c^2_{xy}	c^2_{yx}
Real values	500	550	250	260	550	560	260	270
Calculated values	506.9	547.2	259.7	259.5	557.0	558.6	268.7	271.9
Relative error %	1.38	0.51	3.88	0.19	1.27	0.25	3.35	0.70
	Mass coefficients (kg)							
	m^1_{xx}	m^1_{yy}	m^1_{xy}	m^1_{yx}	m^2_{xx}	m^2_{yy}	m^2_{xy}	m^2_{yx}
Real values	—	—	—	—	—	—	—	—
Calculated values	2.364	2.430	−0.069	−0.028	2.341	2.284	−0.027	−0.055
Relative error %	—	—	—	—	—	—	—	—

does not exceed 0.8%. The mass of the shaft was determined with an accuracy of 0.012%. The coefficients on the main diagonals were determined with greater accuracy than those in cross-coupled directions.

7.8 Summary

The sensitivity analysis of the method of determination of 24 dynamic coefficients of bearings is presented in this chapter. In order to verify the parameters, a numerical model in the Samcef Rotors software was used. The model consisting of a rotor and two bearings made it possible to change parameters that were difficult or impossible to verify during experimental tests.

As part of the work, calculations were carried out for the reference rotor model – a symmetrical rotor with a disk in the center and bearing supports placed at equal distance to the center. Since the values of stiffness and damping coefficients as well as the mass of the shaft were known, it was possible to directly compare the calculated values with those set in the numerical model. After calculations based on the reference model, stiffness coefficients are obtained with an error of up to 0.63%. Damping coefficients are calculated with an error of up to 3.76%. The values of stiffness and damping coefficients on the main diagonals are calculated with double the accuracy in this case. The mas of the shaft can be calculated with very high accuracy of up to 0.003%. This accuracy makes it possible to calculate the mass of the shaft, which is 4.845 kg, with an accuracy of 0.0001 kg.

One of the checked parameters was the displacement of the excitation force from the initial center of the rotor by 30 mm next to the disk. It turns out that even a displacement as small as 5% of the rotor length between bearings results in calculation error of about 12% for stiffness coefficients, of about 15.5% for damping coefficients and 0.2% for the determined mass of the shaft. The accuracy of the calculation was improved by dividing the force per bearing proportionally. This operation resulted in a reduction of the resulting calculation error to a level close to the reference model. In this case, it was also interesting to compare the calculated mass coefficients with their expected value (half the mass of the shaft). In the case without asymmetrical force distribution, the error in calculating these coefficients was about 15%, while the proportional distribution of force on both bearings by the application of an appropriate force distribution reduced its value to about 3.8%. On the basis of the values of these coefficients, already at the initial stage of research, the correctness of determining the set of dynamic coefficients of bearings can be assessed. This example shows the rationale for calculating in a single operation the mass coefficients in addition to the stiffness and damping coefficients, which are the searched values.

If we define the force in the calculation algorithm correctly, we will get the correct values of dynamic coefficients of bearings. During experimental tests there is the possibility of an "unclean" impact. The force generated by the impact hammer will not be transmitted in one direction only. An experiment was carried which consisted in rotating the force F_x by 15°. This is quite a large angle, but it is worth checking what error this operation produced. The calculation of dynamic coefficients of bearings was carried out as if the excitation was made at an angle of $\alpha = 0°$. Such an approach may result in a calculation error as high as 56%. This error occurs only for some coefficients. The largest calculation error occurred when calculating *YX* cross-coupled coefficients. A significant error was also generated during the calculation of *YY* coefficients. Other coefficients are determined with an accuracy of up to 3%. If the force excitation is correct, this error will not occur during experimental research.

In experimental tests, it is usually impossible to directly measure the rotor displacement in bearings, and it is required for calculations. Dynamic coefficients of bearings were calculated basing on a signal measured at a distance of 30 mm from the center of the bearing. It turns out that appropriate geometric transformation of the measured signals ensures that correct values of dynamic factors are obtained. Failure to take this displacement into account will result in calculations with an error of approximately 2%.

Since a dynamic test is performed during the experimental research, consisting of a two-fold excitation of the spinning rotor with a impact hammer, the influence of the rotor material stiffness on the results of calculations was checked as well. All previous calculations were made for a rotor with Young's modulus value (210 GPa) equal to that for steel. A 100-fold increase in the modulus of longitudinal stiffness of the material was analyzed. After this change, the same results for dynamic coefficients of bearings were obtained. This means that the rotor material does not affect the calculated values.

In order to verify the results of calculation for an asymmetrical rotor, the basic numerical model of the rotor was modified. In this analysis one side of the rotor was 60 mm shorter than the other. Such cases occur when, for example, it is necessary to mount a coupling on one side. Calculations of dynamic coefficients of bearings made for such a rotor can be considered correct. The values of calculation errors were similar to those obtained from the reference model.

Some parameters have no impact on the results of the calculations, whereas others do and it is difficult to correct them later. It was shown that certain parameters, such as displacement of excitation force or displacement of measuring sensors, result in measurement error and their negative impact on results can be compensated for with appropriate corrections. Such information may be helpful in the assessment of the calculation error that may be caused by several of the examined factors simultaneously. The guidelines provided in this chapter are helpful in the experimental determination of dynamic coefficients of bearings, and their detailed description shows the possibilities and limitations of this method.

8

Experimental Studies

The method described in the three previous chapters was used to calculate dynamic coefficients of bearings on the basis of experimental research. Chapter 5 presented the calculation method, Chapter 6 presented one of the steps of signal processing, and Chapter 7 described the sensitivity analysis of the method. The information presented in these chapters provides guidelines on conducting experimental research.

This chapter describes experimental tests, their preparation, measured signals, dynamic coefficients of hydrodynamic bearings calculated on this basis and verifications of results. Presented at the beginning of this chapter are two programs written in Matlab language equipped with a graphical user interface (GUI), which was created with the use of GUIDE tools. The first program was called "Signal;" its description is provided in Section 8.1. It is a simple program for preparing signals obtained during experimental research for further calculations. Its operation consists in selecting from the time course of fragments of the rotor displacement signal (after inducing the rotor with an impact hammer), then a reference signal (stable operation) is selected and on the basis of these two signals a third signal is generated, which is generated by subtraction of the previous two signals. The operations are carried out in accordance with Figure 6.1 in Chapter 6. The second program, called "Calculation," the detailed description of which is provided in Section 8.2, was used for calculations of dynamic coefficients of bearings. This program implements the algorithm described in Chapter 5. The purpose of this algorithm is to determine the stiffness, damping, and mass coefficients of the rotor–bearing system.

In Section 8.3 the activities necessary for the preparation of the test rig, i.e. alignment and balancing of the rotor, are described. These preparations contributed to the reduction of initial vibrations, and thus to a more stable operation of the system. The analysis of excitation of the rotor with an impact hammer by means of a high-speed camera was also carried out. These studies allowed the slippage of an impact hammer during the excitation of a rotating shaft to be observed.

Section 8.4 presents the characteristics of the system during stable operation, the course of the excitation force and the dynamic responses recorded after excitation with an impact hammer. Based on the data collected in this way, it was possible to calculate dynamic coefficients of bearings. For the calculation it was also necessary to consider the actual position

Bearing Dynamic Coefficients in Rotordynamics: Computation Methods and Practical Applications,
First Edition. Łukasz Breńkacz.

Companion website: www.wiley.com/go/brenkacz/bearingdynamiccoefficients

of the rotor journal displacement sensors, which are shifted by 30 mm in relation to the center of the bearing supports. It was also taken into account that the excitation force was not applied exactly in the middle of the rotor length, but next to the disk. Also, a transformation that takes into account the fact that eddy current sensors are rotated at an angle of 45° to the global reference system was presented. Operations carried out on signals obtained from experimental research are presented in Section 8.5.

On the basis of appropriately prepared signals, mass, stiffness and damping coefficients of the rotor–bearing system were calculated. The results of mean and standard deviation values are presented in Section 8.6.

Section 8.7 provides a verification of the calculated stiffness and damping coefficients. A model consisting of a material point with one degree of freedom which was connected to another fixed point by means of an elastic-damping element was created in the Abaqus program. The stiffness and damping of the connecting element were determined on the basis of the results of experimental tests. The impulse force measured during experimental studies was used to induce vibrations, and then the response of this system was recorded. The positive verification of the calculated stiffness and damping parameters in this case was evidenced by the conformity of the response measured during the experimental tests and that calculated based on the estimated stiffness and damping coefficients. Section 8.8 provides a summary and conclusions.

8.1 Software Used for Processing of Signals from Experimental Research

In order to carry out the process of calculation of dynamic coefficients of bearings, it was necessary to prepare the signal properly. During the experimental tests, 40 seconds of operation was measured for each rotational speed. The signal from four eddy current sensors and an impact hammer was recorded. During this time, the rotor was induced in one direction (*X* or *Y*) using an impact hammer. Depending on the case, it was about 25 excitations per measurement. Then this operation was repeated in the other direction. In order to prepare the signal, a program called "Signal" was created. Extracts from the code of this program together with descriptions are presented in Appendix B.

The "Signal" program was equipped with a GUI. The application window consists of seven parts numbered 1–7 (Figure 8.1). Using the buttons denoted by 1, it was possible to load the signal from experimental tests. Force courses from one direction (*X* or *Y*) and system response signals from four eddy current sensors from two directions (*X* and *Y*) were loaded. In one operation a signal was prepared for excitation in the *X* or *Y* direction, then the process is repeated for the other direction. Using the checkboxes found in the part numbered 2 it was possible to select the signals to be shown in the diagrams. It is possible to display one, two, three, or four displacement signals and any combination of them. These checkboxes serve as a filter. The possibility of quick verification of only some of the courses proved to be extremely helpful during signal analysis. The buttons for refreshing the diagrams are denoted by 3. They were also activated automatically after each change of signal range and filters. The range of the signal for analysis was selected by means of sliders and text boxes located in the part numbered 4. These elements determined (in this order):

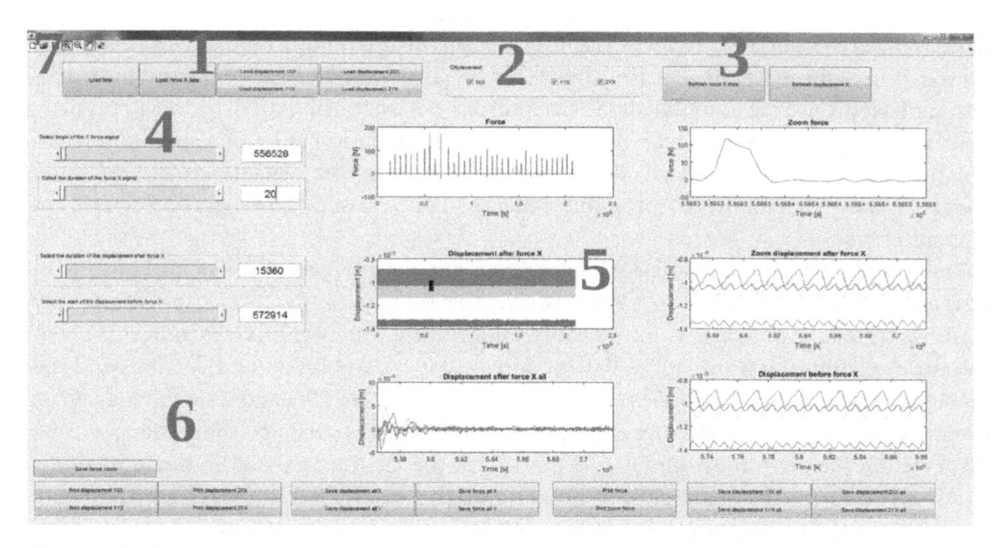

Figure 8.1 Program window for signal preparation; the diagrams presented the data read for a signal corresponding to a speed of 4500 rpm and excitation in the *Y* direction.

- The beginning of the excitation force signal.
- Duration of the excitation force.
- The beginning of reference signal.
- Duration of the reference signal.

Signals from experimental research are presented in the diagrams in the part numbered 5. The diagrams show in order (in rows):

- Excitation force signal over the entire time range.
- Excitation force signal (interval specified in the part numbered 4).
- System response signal over the entire time range.
- Signal after excitation (interval specified in the part numbered 4).
- The signal used later in the calculation (the reference signal was subtracted from the signal after excitation).
- Reference signal (interval specified in the part numbered 4).

In the part numbered 6 there is a set of buttons which are used to display windows with enlarged fragments of previously described signals and to save them. The part numbered 7 features tools which help perform operations on diagrams. They enable zooming in and out of fragments of the signal, moving it and checking the coordinates of points on the diagrams. When changing the size of the window, all elements of the interface are scaled.

8.2 Software Used for Calculations of Dynamic Coefficients of Bearings

A program called "Calculation" was developed to calculate dynamic coefficients of bearings. The heart of this program is the algorithm described in Chapter 5. Additionally, the program was enhanced with operations of loading files, saving the results and displaying data at different stages of calculation. It was used for operations on a signal previously prepared in Samcef Rotors – during algorithm verification (Chapters 4–6) and for calculations based on experimental data (Chapter 7). The data prepared in the previously described "Signal" program were loaded into the "Calculation" program. Before the calculations, the operations described in Section 8.5 were also performed. Extracts from the code of the "Calculations" program together with descriptions are presented in Appendix B.

The interface of the "Calculation" program is shown in Figure 8.2. The program window can be divided into 22 parts. Part 1 allows to read data in the X direction, while Part 2 allows to read input signals in the Y direction. In the calculations it was necessary to take into account the signals of excitation forces in two perpendicular directions and the response of the system to these excitations. Parts 3 and 4 are checkboxes, which acted as filters. They made it possible to choose which response signals of the system will be displayed in the diagrams numbered 8 and 12 for the X direction and in diagrams numbered 9 and 13 for the Y direction. The button denoted by the number 5 is responsible for refreshing the diagrams. Refreshing is also performed automatically after each change of parameters and filters. Diagram 6 shows the excitation force signal in the X direction, while diagram 7 shows the excitation force signal in the Y direction. The curves of excitation forces after the fast Fourier transform (FFT) analysis (in the frequency domain) are presented in Parts 10 and 11. Diagram 8 shows the response signal of the system after excitation in the X direction, while diagram 9 shows the response signals of the system in the Y direction. Diagrams of this signal after the FFT analysis are presented in Parts 12 and 13. Part 14 features a diagram of dynamic stiffness, which is created by multiplying the excitation force and the response of the system. A diagram

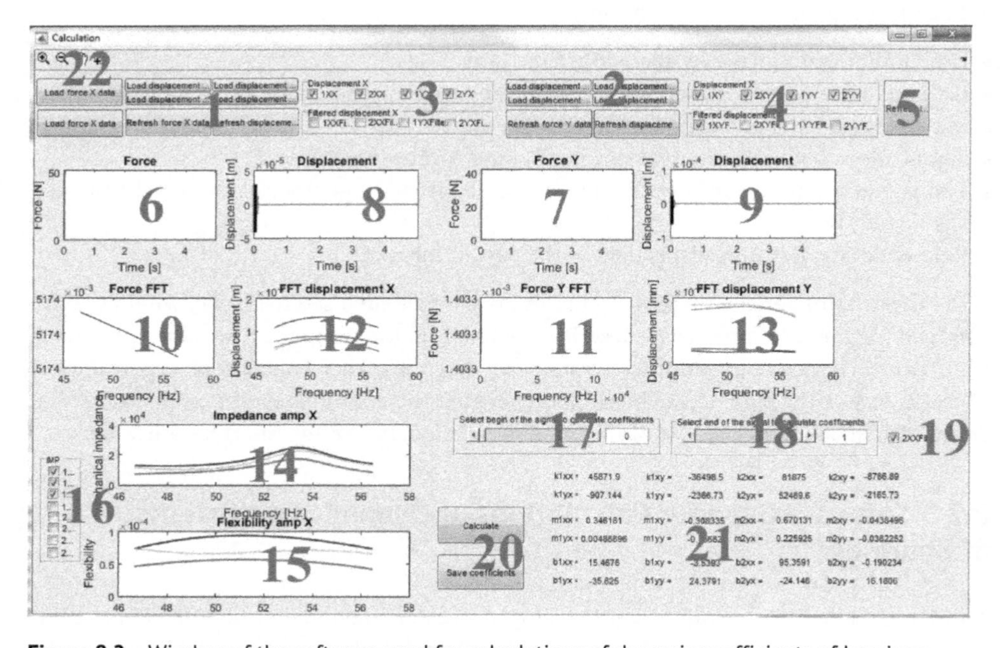

Figure 8.2 Window of the software used for calculations of dynamic coefficients of bearings.

of dynamic flexibility is presented in Part 15, which is the inverse of the flexibility signal. Part 16 contains "filters," which allowed to specify which diagrams of stiffness and dynamic flexibility were displayed. Sliders and text fields in Parts 17 and 18 make is possible to define the signal length, which is used to calculate dynamic coefficients of bearings. This restriction was taken into account by ticking the checkbox of the type shown in Part 19. After marking this field vertical black lines symbolizing the signal range taken for further calculations were visible in diagrams 8 and 9. In Part 20 there are buttons used to calculate and save dynamic coefficients of bearings. The button for calculating dynamic coefficients of bearings was engaged automatically when the calculation parameters were changed, so that it was possible to quickly verify how dynamic coefficients of bearings are affected by the introduced changes in parameters. The results of the calculated coefficients are presented in Part 21. The option of saving dynamic coefficients of bearings to files, made it possible to directly compare them and to draw up the course of changes in their values as a function of the rotational speed. The upper part of the program window denoted by 22 features buttons that can be used to perform operations on diagrams 6–15. They make it possible to increase, decrease, and move each of the diagrams. Additionally, it is possible to check the coordinates of points on the diagrams.

8.3 Preparation of Experimental Tests

Before the experimental tests were performed, it was necessary to prepare a laboratory test rig. The process of alignment of the motor shaft and the shaft of the laboratory test rig using plain bearings was carried out. This operation should be carried out prior to the first run of almost any rotating machinery. The process of balancing the rotor was also carried out. As a result of these two operations, the level of vibrations measured on the laboratory stand was reduced several times. The basic characteristics presented in Chapter 3 have been drawn up after the alignment and balancing processes. As part of the preliminary tests, the analysis was also carried out by means of a high-speed camera. This is the best available method for observing phenomena related to the slippage of an impact hammer when inducing the rotor.

Unbalance and misalignment is one of the most common sources of vibration in rotating machinery (Muszyńska 2005). The shaft alignment was carried out using OPTALIGN Smart RS from Prüftechnik and two wireless laser sensors. The first laser sensor was mounted on the shaft of the drive motor, while the second one was mounted on the end of the shaft supported by two plain bearings. The shaft alignment tool and the shaft-mounted sensors are shown in Figure 8.3. The dimensions of the machine to be aligned were entered into OPTALIGN Smart RS. The drive motor was a fixed part, while the laboratory test rig baseplate with bearing supports were the adjustable elements. The effect of alignment was a deviation of dimensions at the feet not exceeding 0.01 mm. According to the PN-ISO 10816-1 (1995) standard, it can be stated that the rotor is aligned correctly within the tested speed range and the level of vibrations caused by misalignment is acceptable (Orłowski and Słowański 1978; Niezgodziński and Niezgodziński 2004). The alignment report is provided in Appendix A.

Figure 8.3 Photograph showing the test stand during the shaft alignment.

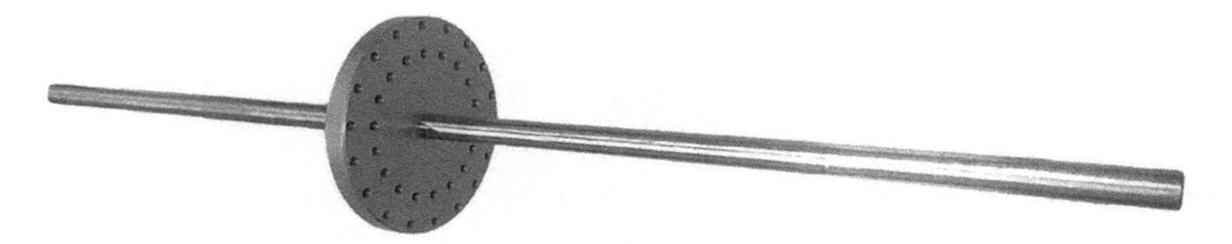

Figure 8.4 Photograph showing the tested rotor.

After the alignment process, the rotor in plain bearings was balanced. The rotor is balanced using a device called Diamond 401 manufactured by MBJ Electronics. It is shown in Figure 4.7. A schematic view of the rotor is shown in Figure 5.1, and a photograph is shown in Figure 8.4. In the center of the rotor there was a disk with a diameter of 6″ (152.4 mm) with 36 threaded holes, which made it possible to balance the rotor. The single-plane balancing consisted of screwing in additional load in the threaded holes in the form of screws. After the balancing process, the residual unbalance of the rotor was about 10 g-mm. According to the ISO 1940-1:2003 (2003) standard this is a permissible value.

The method of inducing the rotor with an impact hammer was important from the point of view of accuracy of the results. During the verification of the sensitivity of the method of calculating the dynamic coefficients of rotating machinery presented in Chapter 7, the influence of the change in the angle of inclination and the change in the place of application of the excitation force on the accuracy of the obtained results was checked. It is not easy to take into account the effect of slippage when inducing the rotor during numerical testing. It is also extremely difficult to observe this phenomenon during experimental research. Eddy current sensors were used to record shaft movements near bearings in two directions perpendicular to each other and to the shaft axis. On this basis, it is not possible to determine whether the excitation with an impact hammer was exactly as intended – in the

Figure 8.5 Phantom Camera Control, a program window for operating a high-speed camera; a fragment of the recording of inducing the rotor with an impact hammer can be seen in the background.

specified direction and without slippage. For this reason, observations were made with the use of the Phantom v2512 high-speed camera manufactured by Vision Research.

Figure 8.5 shows the Phantom Camera Control window used to operate the high-speed camera and basic analysis of movements viewed in slow motion. The aluminum disk can be seen on the left of the photograph. A black-and-white strip of tape was applied to the rotor shaft for precise tracking of the rotor movement. The upper part of the photograph shows the tip of the impact hammer.

After the tests with the use of a high-speed camera it was found that the slippage at the moment of the excitation of the shaft is small and can be ignored. The force transmitted by an impact hammer to the rotor can be treated as an excitation along one of the axes (X or Y). This is very valuable information, which confirms that the recommendations formulated after the numerical sensitivity analysis of the method, presented in Chapter 7, are met.

8.4 Implementation of Experimental Research

The experimental studies were similar to the numerical studies described in Chapters 5 and 6. The calculation scheme is shown in Figure 6.1. Compared with the previous analysis carried out with the use of numerical data, it was extended by an additional step (Figure 8.6).

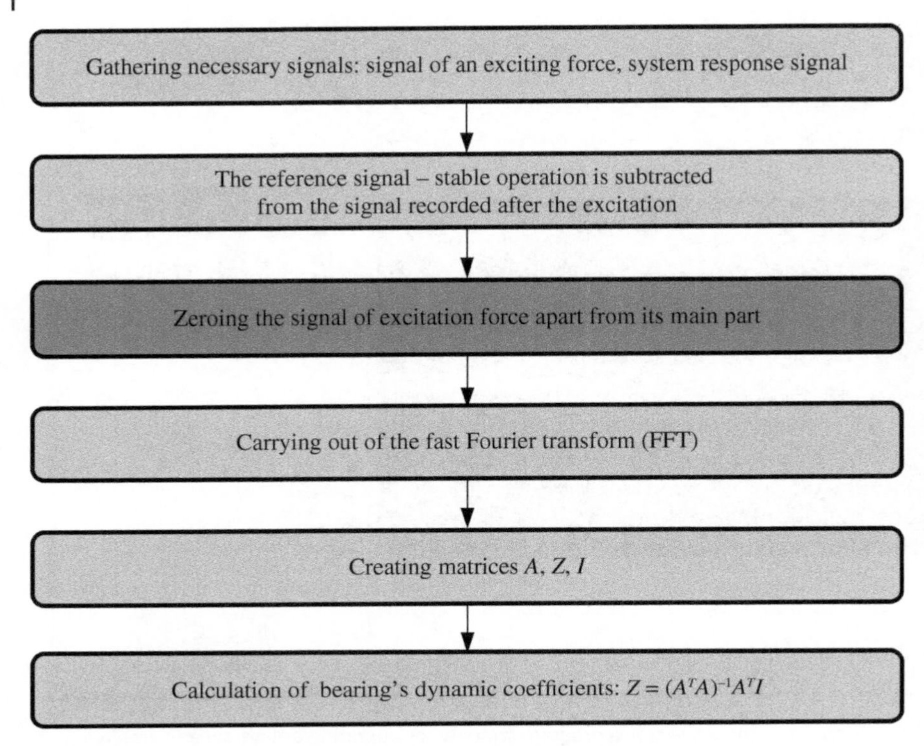

Figure 8.6 Calculation scheme of dynamic coefficients of bearings.

Experimental research is described in more detail elsewhere (Breńkacz et al. 2017a). In the third step, the excitation force signal is reset to zero outside the main component. In the interpretation of the results, analyses presenting basic characteristics of the laboratory test rig proved to be helpful. The results of the analysis of structural dynamics are presented in Chapter 3 and Appendix A.

All measurements were carried out using the Scadas Mobile analyzer, Test.Lab software using eddy current sensors, accelerometers, and a laser tachometer. Figure 8.7 shows the signal measured by an eddy current sensor located near bearing no. 2 in the *Y* direction at a speed of 4500 rpm. During the tests, the rotor was induced in the central part by means of an impact hammer in the *Y* direction. The diagram shows the signal of stable operation of the rotor with a periodic increase in amplitude caused by excitation with an impact hammer. Then this operation was repeated for excitation in a perpendicular direction. Such a procedure was used for each rotational speed tested.

In order to obtain the complete set of data required to calculate the dynamic coefficients of bearings for one rotation speed, it was necessary to carry out the analysis in two stages. In the first stage the rotor was induced in the *Y* direction (in the central part denoted by P_W in Figure 3.2) by means of an impact hammer. In the second stage, the measurement was carried out at the same rotational speed, with excitation in the perpendicular direction, i.e. in the *X* direction. The excitation force signal in the *Y* direction is shown in Figure 8.8. Similar to the displacement signal, this one was also measured for a duration of 40 seconds;

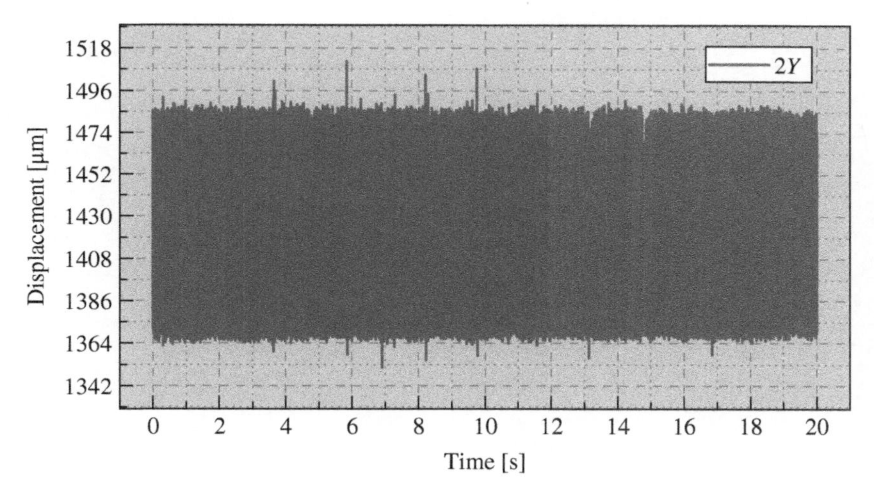

Figure 8.7 Signal measured near bearing no. 2 in the *Y* direction at 4500 rpm and excitation with an impact hammer in the *Y* direction.

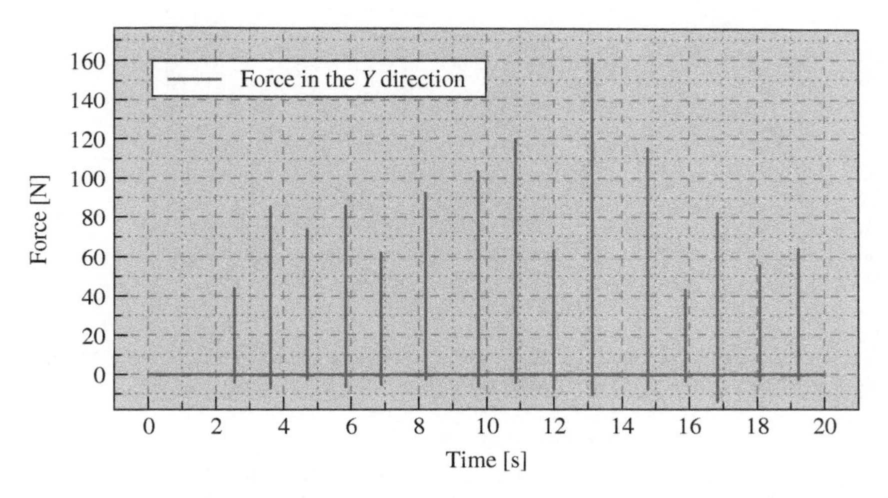

Figure 8.8 Excitation force curve in the *Y* direction over 20 seconds; signal stored at 4500 rpm.

Figure 8.8 shows a 20-second fragment. Signals of excitation force and system responses measured in this way were used to calculate dynamic coefficients of bearings.

The force values varied from 60 to 120 N (usually about 80 N). The excitation force over 0.4 ms is shown in Figure 8.9. The duration of most excitations ranged from 0.1 to 0.2 ms. As each force signal was different, a FFT was performed for each excitation force in the calculations. Values of excitation force outside the main "peak" were reset to zero. The result of this operation is shown in Figure 8.9b.

Figure 8.10a and b shows the signal measured at stable operation of bearing no. 2 in the *X* and *Y* directions. Figure 8.10c shows the signal measured by the eddy current sensor in the *Y* direction after inducing the rotor with an impact hammer in the *Y* direction (the signal of

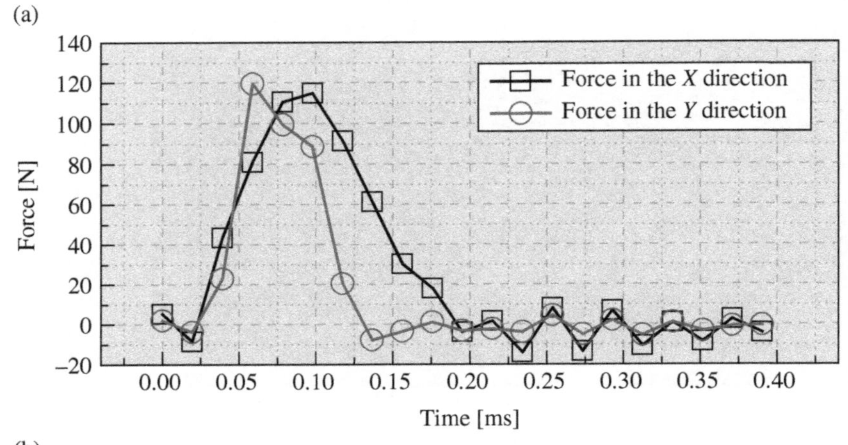

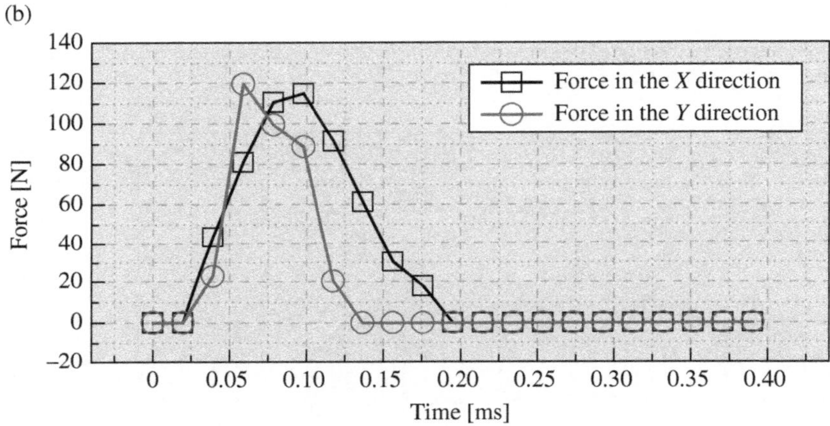

Figure 8.9 Excitation force in the X and Y directions over 0.4 ms; excitation at 4500 rpm. (a) Values not reset outside the main peak and (b) signal with resetting.

this excitation force is shown in Figure 8.9). Figure 8.10d shows the signal generated by the rotor in the X direction. Excitation in this case was also given in the Y direction. You can see that the vibration trajectories of the rotor change after the excitation and then they are stabilized – they return to the stable operation. The amplitude of the signal increased or decreased (for a short period of time) depending on the current location of the rotor.

The "Signal" program described in Section 8.1 was used to select appropriate fragments of reference signals (after excitation) and excitation force. Selecting the appropriate signal ranges was the most time-consuming part of the calculation of dynamic coefficients of bearings. The result of the subtraction operation from the signals shown in Figure 8.10c and d of the signals shown in Figure 8.10a and b is the signal shown in Figure 8.11. Figure 8.11a shows the response of the first bearing after inducing in the Y direction ($1YY$ and $1XY$) and in the X direction ($1XX$ and $1YX$). Similarly, the signals for the second bearing are shown in Figure 8.11b. In the key, the first letter indicates the direction of the displacement measurement and the second letter indicates the direction of the excitation force. The system response signal for the second bearings in the X direction after excitation in the Y direction is shown as $2XY$ in the diagram.

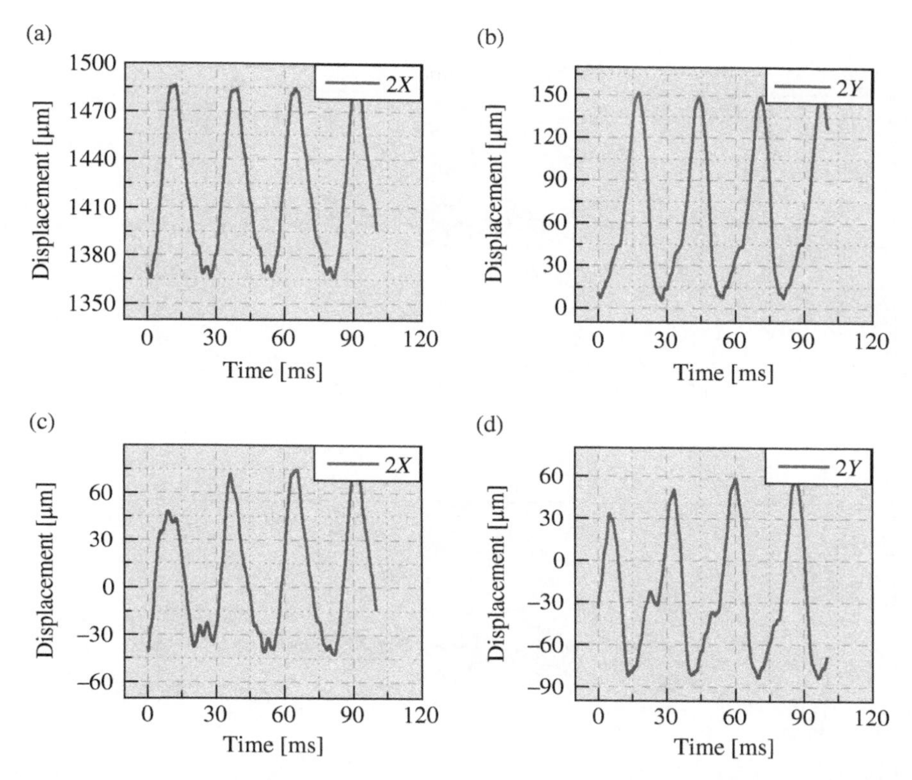

Figure 8.10 Stable bearing operation no. 2 (a and b) at 4500 rpm and signal after excitation (c and d); the graphs show the displacements in the *X* direction (a and c) and *Y* direction (b and d); the time of the signal shown in the graphs is 0.1 seconds.

8.5 Processing of the Signal Measured During Experimental Tests

In order to obtain the expected results, the experimental signal must be processed in an appropriate manner. It is necessary to take into account the fact that eddy current sensors are placed at an angle of 45° to the global reference system. The fact that the signal measured from eddy current sensors is "shifted" compared with the signal that would be obtained if it was measured exactly halfway across the width of the bearing support is also important. It was also necessary to divide the excitation force asymmetrically because it was applied next to the rotor disk. The rotation of the coordinate system for the first bearing was carried out using Eqs. (8.1) and (8.2). An analogous operation must be carried out for the second bearing:

$$\begin{bmatrix} 1XX' \\ 1YX' \end{bmatrix} = \begin{bmatrix} \cos\emptyset & \sin\emptyset \\ -\sin\emptyset & \cos\emptyset \end{bmatrix} \begin{bmatrix} 1XX \\ 1YX \end{bmatrix} \tag{8.1}$$

$$\begin{bmatrix} 1XY' \\ 1YY' \end{bmatrix} = \begin{bmatrix} \cos\emptyset & \sin\emptyset \\ -\sin\emptyset & \cos\emptyset \end{bmatrix} \begin{bmatrix} 1XY \\ 1YY \end{bmatrix} \tag{8.2}$$

(a)

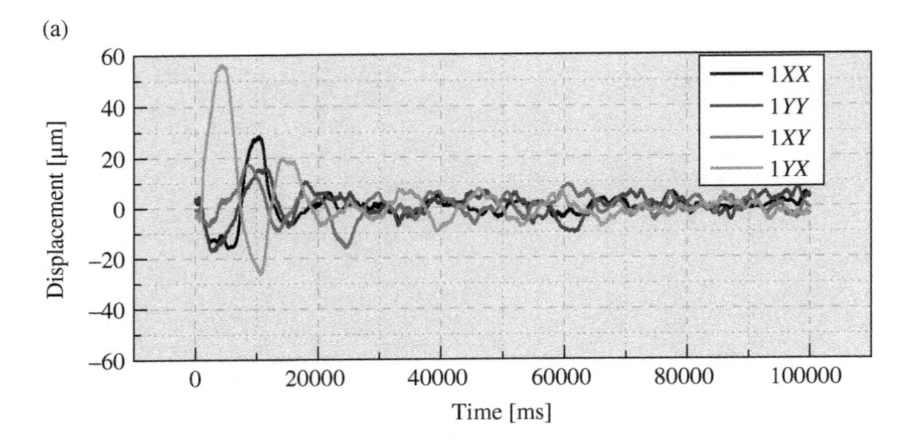

(b)

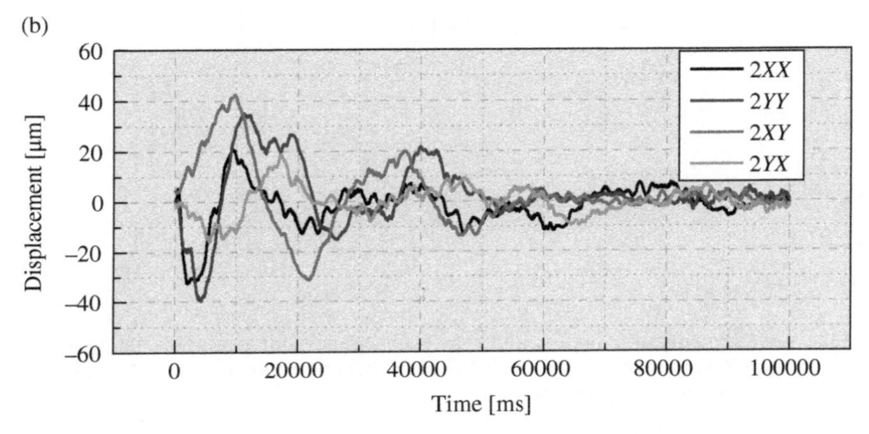

Figure 8.11 Amplitude of rotor vibrations after excitation in the *X* and *Y* directions for the first (a) and second (b) bearings; the reference signal has been subtracted from the signal after the excitation.

The code in the Matlab program used for this operation is as follows:

```
% Rotation of eddy current sensor signal to global coordinate system
alpha = -pi()/4; % [rad] Angle by which the system is rotated
sxx1 =  data1XX*cos(alpha)+data1YX*sin(alpha);
syx1 = -data1XX*sin(alpha)+data1YX*cos(alpha);
sxy1 =  data1XY*cos(alpha)+data1YY*sin(alpha);
syy1 = -data1XY*sin(alpha)+data1YY*cos(alpha);
sxx2 =  data2XX*cos(alpha)+data2YX*sin(alpha);
syx2 = -data2XX*sin(alpha)+data2YX*cos(alpha);
sxy2 =  data2XY*cos(alpha)+data2YY*sin(alpha);
syy2 = -data2XY*sin(alpha)+data2YY*cos(alpha);
```

The signal from eddy current sensors was corrected using Eqs. (7.4) and (7.5). The purpose of this procedure was to correct the fact that eddy current sensors were not located in the center of the bearing, but next to the bearing supports. Based on the sensitivity analysis carried out in Chapter 7, it can be concluded that failure to take this correction into account may result in a calculation error of several percent for dynamic coefficients of bearings. The code in the Matlab program used for this operation is as follows:

```
% Signal shift from sensors to bearings
s1 = 0.260; %m Distance from center of rotor to sensors at
bearing no. 1
s2 = 0.260; %m Distance from center of rotor to sensors at
bearing no. 2
l1 = 0.290; %m Distance from center of rotor to support 1
l2 = 0.290; %m Distance from center of rotor to support 2
xx1 = (1/(s1+s2))*((s2+l1)*data1XX+(s1-l1)*data2XX);
xx2 = (1/(s1+s2))*((s2-l2)*data1XX+(s1+l2)*data2XX);
yy1 = (1/(s1+s2))*((s2+l1)*data1YY+(s1-l1)*data2YY);
yy2 = (1/(s1+s2))*((s2-l2)*data1YY+(s1+l2)*data2YY);
xy1 = (1/(s1+s2))*((s2+l1)*data1XY+(s1-l1)*data2XY);
xy2 = (1/(s1+s2))*((s2-l2)*data1XY+(s1+l2)*data2XY);
yx1 = (1/(s1+s2))*((s2+l1)*data1YX+(s1-l1)*data2YX);
yx2 = (1/(s1+s2))*((s2-l2)*data1YX+(s1+l2)*data2YX);
```

The calculation also takes into account the fact that the excitation force was not applied exactly in the middle between the supports, but at the point P_W, 30 mm from the center of the rotor disk. This fact was taken into account by applying the correction described by Eq. (7.3). The code in the Matlab program used for this operation is as follows:

```
% Asymmetrical excitation force distribution
l = 290; % mm Half of the rotor length
w1 = l - 30; % Distance from first support to excitation
point, 260
w2 = l + 30; % Distance from second support to excitation
point, 320
data1HX = dataHX*w1/l;
data1HY = dataHY*w1/l;
data2HX = dataHX*w2/l;
data2HY = dataHY*w2/l;
```

8.6 Results of Calculations of Dynamic Coefficients of Hydrodynamic Bearings on the Basis of Experimental Research

When calculating mass, stiffness and damping coefficients of the rotor–bearing system, the experimental data described in Section 8.4 were used. Selection of appropriate fragments of signals and their initial processing was done with the use of the "Signal" program described in Section 8.1. Data were loaded into the "Calculation" program described in Section 8.2.

Using this program, calculations of mass, damping and stiffness coefficients of the rotor–bearing system were carried out.

In the first step, after loading the data into the "Calculation" program, the operations described in Section 8.5 were performed. In the next steps, the procedure described in Chapter 5 was implemented. A set of 24 dynamic coefficients of bearings for each rotation speed was calculated on the basis of at least 10 data sets. Then four sets of data were rejected, resulting in the highest value of standard deviation. The mean value (3.2) and standard deviation (3.3) for 24 dynamic coefficients of bearings were calculated on the basis of six sets of data. Calculations were made for the entire speed range. A summary of the stiffness, damping and mass coefficients of the rotor–bearing system is shown in Table 8.1. Table 8.2 lists the standard deviations of all coefficients. The results are described in more detail elsewhere (Breńkacz et al. 2017b).

The list of stiffness, damping and mass coefficients is presented in a graphical form for the second bearing only (the one further away from the coupling). The results for the first bearing were less accurate due to the influence of the fixed coupling. The stiffness coefficients of the second bearing are shown in Figure 8.12. Since bearing no. 2 is at a distance from the impact of the coupling system, the results of the dynamic coefficients calculated for it will be directly compared with the numerical calculations in the latter part of this work (Chapter 9). The results in all diagrams in this chapter can be divided into two parts: the first in the speed range from 2250 to 3750 rpm; and the second for speed above 3750 rpm. This division can be defined on the basis of the magnitude of the vibration trajectory of the rotor. In the first interval, vibrations of a small amplitude were observed. In the second interval, both vibrations of higher amplitudes and resonance were present. This division is also visible in the diagrams of stiffness coefficients of hydrodynamic bearings. Results of calculations made for speeds above 3750 rpm are characterized by a greater standard deviation.

It is also possible to define a division based on the observation of the vibration spectrum. Up to 2750 rpm, the dominant influence of the synchronous component can be seen in the vibration spectrum generated by the second bearing. Above this speed, the influence of the subsynchronous component dominated. This indicates hydrodynamic instability and non-linear properties of the bearing lubricating film. This phenomenon can be caused by incorrect bearing geometry and radial clearance, inadequate bearing supply pressure, changes in oil viscosity due to temperature changes, influence of flaccid rotor, and bearing housing misalignment.

Figure 8.13 shows a summary of the damping coefficients. As it can be clearly seen, as the speed increases, the values of the damping coefficients decrease – they approach zero. In the vicinity of a rotational speed of 4000 rpm, the cross-coupled damping coefficient $c^2{}_{xy}$ value is negative. This means that vibrations are not suppressed after the excitation, but rather increase their amplitude.

Figure 8.14 shows the results of calculations of the mass coefficients of the second bearing. It is interesting that if the curves defining the coefficients $m^2{}_{xx}$ and $m^2{}_{yy}$ in the second bearing are approximated using a straight line (or a low order polynomial curve), one of the curves increases and the other decreases. They intersect and pass through zero at the rotational speed at which the rotor resonance occurs. If the mass of the shaft is obtained based on calculated mass coefficients, it turns out that this value for rotational speed of 2250 rpm is: $m^2{}_{xx} + m^2{}_{yy} + m^2{}_{xy} + m^2{}_{yx} = 0.3 \pm 0.25 + 0.47 \pm 0.08 + 0.04 \pm 0.09 + 0.37 \pm 0.14 = 1.18 \pm 0.56$ kg. Comparing this value with half the mass of the shaft (i.e. 2.35 kg), the obtained result is about twice as low as expected.

Table 8.1 Stiffness, damping and mass coefficients of the rotor–bearing system for the entire speed range.

	Rotational speed (rpm)										
	2250	**2500**	**3000**	**3500**	**3750**	**4000**	**4250**	**4500**	**5000**	**5500**	**6000**
k^1_{xx} (N/m)	13129.97	11111.59	22418.12	7084.34	17686.06	10108.86	17533.73	43453.03	20777.13	19091.62	34844.18
k^1_{yx} (N/m)	−19404.50	−12682.60	−12960.60	−8718.15	1429.20	−28951.51	−13044.24	24823.69	−6945.56	9761.57	−6471.79
k^1_{xy} (N/m)	−15368.10	−19836.30	7293.66	−3751.26	14772.83	−1084.53	−3951.20	46104.02	8575.60	−20601.50	−11561.90
k^1_{yy} (N/m)	−27452.30	3916.92	10085.29	−529.16	20257.80	11694.55	−7745.56	87801.53	17035.07	13606.88	16948.30
k^2_{xx} (N/m)	3738.66	1839.28	5989.56	−279.32	6632.48	−1987.84	6581.55	9732.91	14277.92	9308.21	28996.16
k^2_{yx} (N/m)	17530.06	15078.57	13801.58	7617.08	7216.56	11741.79	20706.14	20723.56	23366.90	17762.10	10592.05
k^2_{xy} (N/m)	−8415.22	−12499.80	−4917.58	−5019.65	−1754.07	−19983.47	−11586.03	−11892.40	−3590.60	−9320.61	−6720.90
k^2_{yy} (N/m)	11407.26	6429.03	3307.58	294.64	3751.68	20062.97	9692.72	14732.76	1416.96	8689.45	12429.43
m^1_{xx} (kg)	0.18	−0.01	0.14	−0.13	0.22	−0.08	0.15	0.52	0.14	0.06	0.28
m^1_{yx} (kg)	−0.48	−0.28	−0.25	−0.01	0.14	−0.49	−0.17	0.58	−0.07	0.06	−0.09
m^1_{xy} (kg)	−1.10	−0.89	0.31	−0.09	0.30	0.17	−0.12	0.66	0.22	−0.21	−0.15
m^1_{yy} (kg)	−2.29	−0.38	−0.17	−0.19	0.34	−0.13	−0.29	1.31	0.06	−0.01	0.06
m^2_{xx} (kg)	−0.30	0.01	0.12	−0.02	0.12	−0.15	0.11	0.11	0.21	0.08	0.30
m^2_{yx} (kg)	0.37	0.37	0.27	0.04	0.11	0.24	0.31	0.22	0.21	0.20	0.09
m^2_{xy} (kg)	0.04	−0.35	−0.01	−0.08	−0.03	−0.40	−0.19	−0.13	0.06	−0.09	−0.03
m^2_{yy} (kg)	0.47	0.16	0.14	−0.04	0.11	0.28	0.17	0.16	−0.14	0.09	0.09
c^1_{xx} (N·s/m)	19.22	26.48	32.30	37.91	27.17	26.20	29.84	25.36	17.67	32.82	26.79
c^1_{yx} (N·s/m)	−23.67	−50.88	−77.99	−28.46	−23.02	−22.87	−30.33	−53.58	−15.73	−14.09	−7.98
c^1_{xy} (N·s/m)	−37.26	−58.52	7.18	−27.61	6.51	−35.76	−11.74	22.03	8.58	−3.13	−4.08
c^1_{yy} (N·s/m)	−59.37	−39.95	−63.15	−30.10	−21.29	−58.72	−31.09	16.96	−3.61	19.96	5.49
c^2_{xx} (N·s/m)	109.22	59.20	56.47	19.05	39.47	−0.90	33.92	25.00	21.72	4.60	24.84
c^2_{yx} (N·s/m)	16.60	15.03	28.42	14.85	23.78	17.19	5.30	14.89	10.37	8.82	10.14
c^2_{xy} (N·s/m)	27.46	16.08	12.26	−14.08	−5.70	−14.30	−12.53	−4.97	14.50	−3.96	7.24
c^2_{yy} (N·s/m)	67.26	49.34	36.01	13.23	10.44	6.66	20.59	18.83	39.39	12.36	16.87

Table 8.2 Standard deviation of stiffness, damping and mass coefficients of the rotor – bearing system for the entire speed range.

	Rotational speed (rpm)										
	2250	2500	3000	3500	3750	4000	4250	4500	5000	5500	6000
k^1_{xx} (N/m)	18 510.31	8361.37	12 651.45	12 751.96	11 192.71	16 083.35	6083.31	18 307.57	2758.55	7435.09	−14 895.70
k^1_{yx} (N/m)	14 023.57	7251.30	8272.12	14 957.74	14 771.10	39 609.32	5508.85	28 356.51	6085.23	16 875.40	−12 097.70
k^1_{xy} (N/m)	26 721.99	13 165.25	11 635.00	22 561.11	25 084.54	43 737.49	9023.77	48 398.55	20 286.47	27 277.33	−6689.41
k^1_{yy} (N/m)	31 735.00	8795.48	9762.35	35 039.72	33 580.92	16 901.08	20 952.67	64 874.68	19 091.58	11 064.84	−18 835.90
k^2_{xx} (N/m)	2431.75	2158.12	3610.02	3643.25	4145.85	11 527.99	10 262.12	8312.26	8924.33	9720.35	−28 341.40
k^2_{yx} (N/m)	3111.00	16 007.06	7538.59	5965.23	5775.27	4258.53	9820.74	17 592.65	14 655.58	20 419.85	−27 293.20
k^2_{xy} (N/m)	1324.34	5487.81	9423.40	4596.42	4632.32	19 296.52	5920.14	16 332.51	21 680.53	4014.39	−6862.65
k^2_{yy} (N/m)	2225.28	3459.58	7857.16	5179.26	4705.79	25 572.89	15 390.78	14 107.36	20 182.95	13 330.93	−10 100.30
m^1_{xx} (kg)	0.98	0.17	0.38	0.22	0.18	0.62	0.13	0.40	0.05	0.09	−0.17
m^1_{yx} (kg)	0.99	0.51	0.47	0.33	0.31	1.27	0.08	0.55	0.09	0.15	−0.17
m^1_{xy} (kg)	1.48	0.69	0.43	0.53	0.53	1.07	0.20	0.85	0.33	0.23	−0.06
m^1_{yy} (kg)	1.82	0.42	0.48	0.64	0.61	0.37	0.40	1.09	0.42	0.14	−0.24
m^2_{xx} (kg)	0.25	0.14	0.09	0.06	0.07	0.29	0.14	0.10	0.14	0.10	−0.31
m^2_{yx} (kg)	0.14	0.54	0.13	0.12	0.09	0.23	0.11	0.40	0.15	0.20	−0.28
m^2_{xy} (kg)	0.09	0.18	0.24	0.13	0.14	0.44	0.11	0.23	0.29	0.05	−0.06
m^2_{yy} (kg)	0.08	0.22	0.34	0.08	0.12	0.50	0.27	0.13	0.33	0.16	−0.09
c^1_{xx} (N·s/m)	60.27	45.88	46.57	4.68	5.11	22.70	21.16	18.58	19.64	12.59	−9.30
c^1_{yx} (N·s/m)	51.10	21.48	42.56	20.80	17.25	45.30	15.60	36.27	9.88	13.27	−13.78
c^1_{xy} (N·s/m)	81.73	28.97	81.37	29.19	28.12	47.62	22.86	36.66	17.84	18.96	−22.53
c^1_{yy} (N·s/m)	138.10	76.00	82.38	26.30	25.03	81.30	28.81	25.17	42.97	37.87	−42.27
c^2_{xx} (N·s/m)	20.71	33.09	22.22	16.33	18.41	13.66	18.53	14.69	6.82	10.59	−27.78
c^2_{yx} (N·s/m)	7.03	10.11	11.30	8.08	6.44	18.30	17.21	10.77	23.29	13.00	−16.80
c^2_{xy} (N·s/m)	13.89	20.86	13.54	12.71	11.80	33.29	12.63	21.63	17.71	7.36	−23.26
c^2_{yy} (N·s/m)	6.21	21.59	22.08	14.98	15.68	26.68	14.93	17.20	25.46	24.02	−18.82

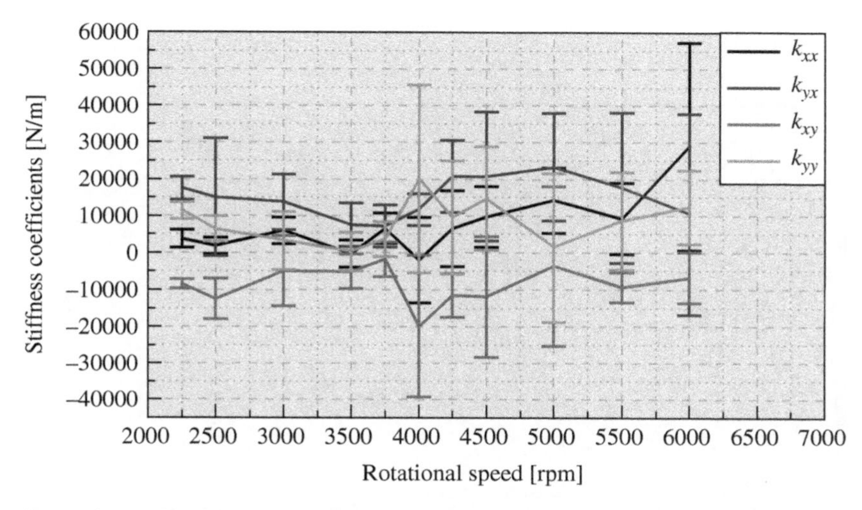

Figure 8.12 Changes in the stiffness coefficients of bearing no. 2 for the entire speed range.

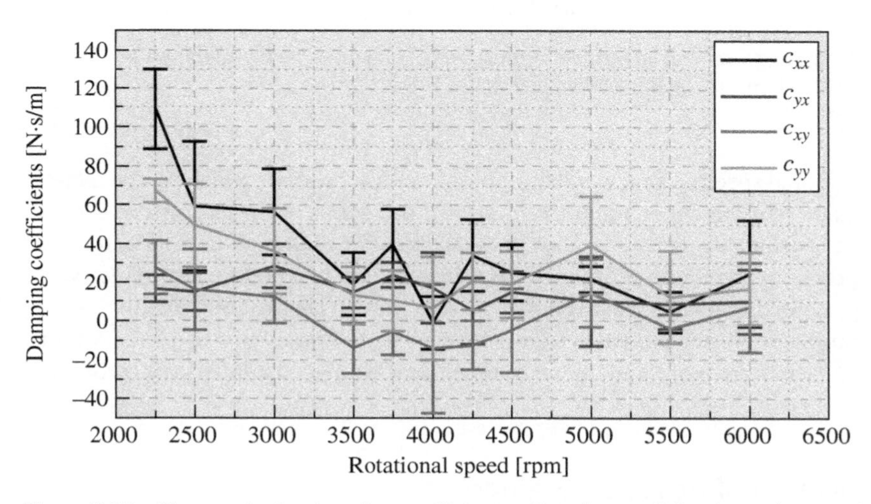

Figure 8.13 Changes in the damping coefficients of bearing no. 2 for the entire speed range.

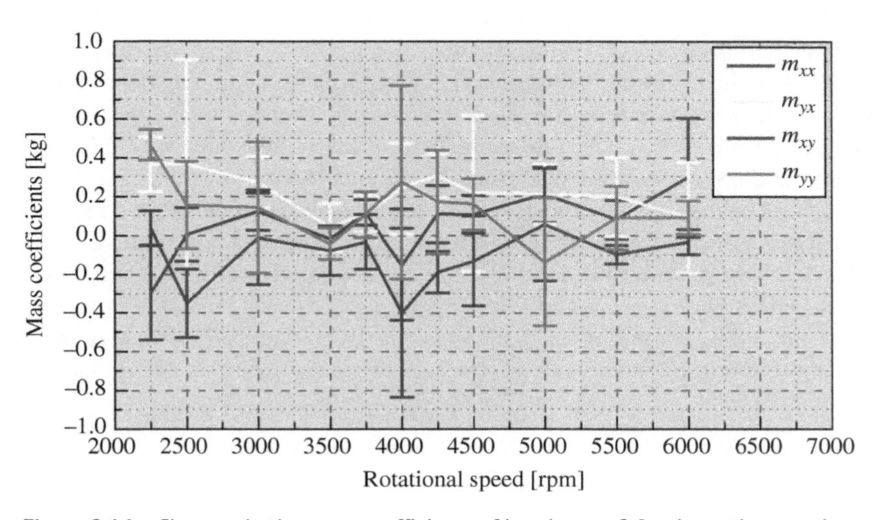

Figure 8.14 Changes in the mass coefficients of bearing no. 2 for the entire speed range.

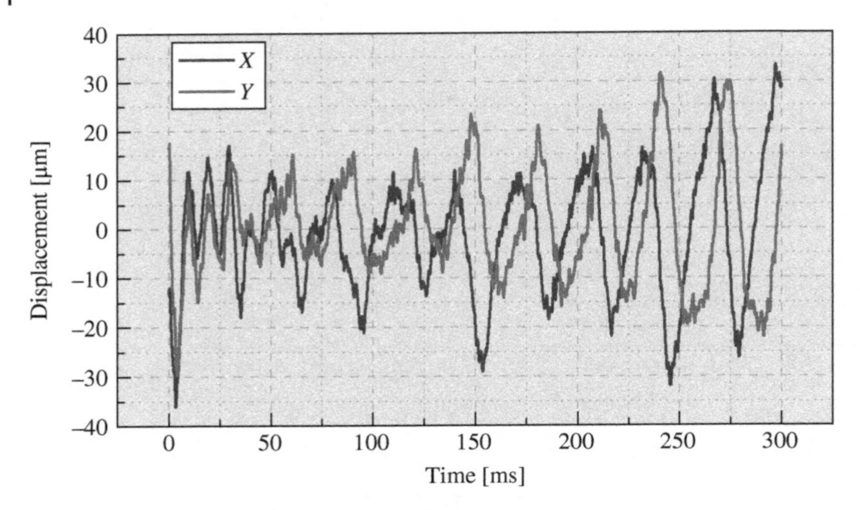

Figure 8.15 Increasing vibration amplitude recorded in bearing no. 2 during operation at 4000 rpm after excitation with an impact hammer in the *Y* direction.

At some rotational speeds a negative damping value was observed, which can be related to the resonance phenomenon. Figure 8.15 shows the signal recorded on the second hydrodynamic bearing during its operation at 4000 rpm after excitation in the *Y* direction. The stable operation signal was subtracted from the excitation signal in accordance with the scheme presented in Chapter 6 (Section 6.1). As it can be clearly seen, during 0.3 seconds the vibration amplitude increases from 0 to more than 30 μm. Such behavior is characteristic of unstable systems and its occurrence can lead to damage to the rotating machine.

8.7 Verification of Results Obtained

In order to verify experimental results, a model consisting of mass of a single degree of freedom, damping and stiffness was developed in the Abaqus 6.14-2 program. In this model the value of force used in experimental tests as well as stiffness and damping determined based on experimental tests have been implemented. The displacement of the mass point (to which the impulse force was applied) determined by means of the model was compared with the response of the system measured during experimental tests.

The model in the Abaqus program consisted of two points lying on one plane. For the first point, all degrees of freedom were removed, and for the second point, one degree of freedom was left – movement in the *Y* direction. A mass equal to half of the mass of the rotor (measured during experimental tests), i.e. 2.35 kg, was assigned to it. Between the two points, an elastic-damping element was inserted. The stiffness value was 8700 Nm, while the damping value was 51 Nm/s. A schematic view of the model is shown in Figure 8.16.

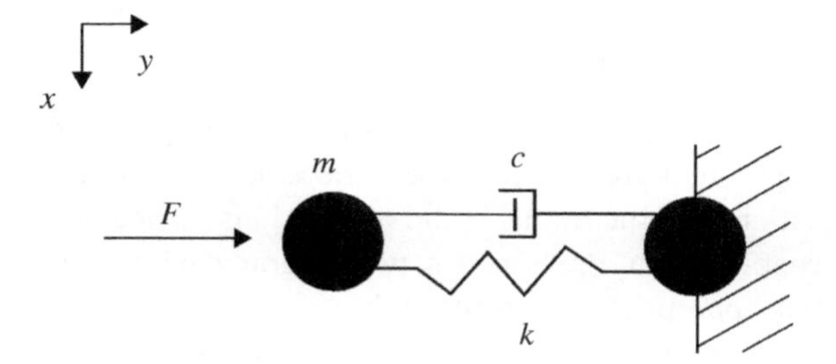

Figure 8.16 Schematic model of mass, damping, and stiffness.

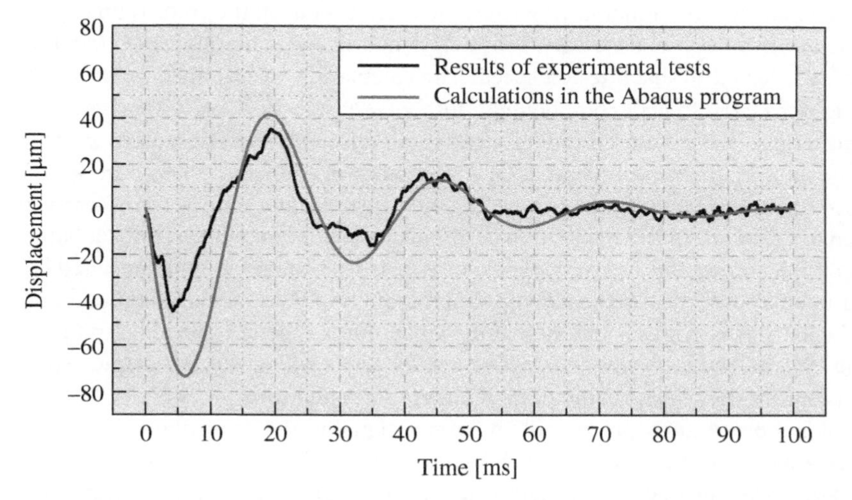

Figure 8.17 Comparison of the dynamic response of the real rotor and the numerical model (the model takes into account the experimentally determined stiffness and damping coefficients).

The value of excitation force was taken as half of the force measured during experimental tests. The force recorded during experimental tests is shown in Figure 8.9b and marked as a force in the Y direction. The values adopted during the verification process accounted for half of this force, i.e. in successive time steps: $F = [0, 0, 11, 59, 49, 44, 10, 0]$. The time step during the analysis was consistent with the frequency of measurement in experimental tests. The duration of excitation force was the same as during experimental tests and was approximately 0.1 second. Calculations were carried out for a time equal to 1 second.

The results of the displacement of a point of one degree of freedom calculated on the basis of the model presented in this subsection are shown in Figure 8.17. The figure also shows the signal measured during experimental tests (this is one of the signals shown in Figure 8.11b). A summary of results confirms that the stiffness and damping coefficients calculated on the basis of experimental tests describe the signals measured during the experimental tests well.

8.8 Summary

This chapter presents experimental tests aimed at determining the stiffness, damping and mass coefficients of the rotor–bearing system. It presents the process necessary to estimate these parameters, the results of calculations and the results of the verifications carried out.

In order to calculate the dynamic coefficients of the bearings, two programs with GUIs were developed. In the "Signal" program appropriate fragments of the signal were selected. Forty-second periods of rotor operation were measured during which vibrations in the *Y* direction were induced a few times with an impact hammer. Then this operation was repeated, but this time the vibrations were induced in a perpendicular direction (*X*). For the calculation, fragments of the signal lasting from the excitation until the rotor returned to its original operating cycle were necessary. Ten such signals in the *X* direction and 10 in the *Y* direction were selected for each tested speed. The calculations were made for 11 rotor speeds ranging from 2250 to 6000 rpm, giving a total of 220 sets of displacement courses (measurement series). In each measurement series (with four eddy current sensors), vibrations near two supports in two perpendicular directions were recorded. The number of signals included in the calculation had to be doubled as the same number of reference signals needed to be found. In the calculations, 220 runs of excitation with an impact hammer were also used. During this analysis, 1980 characteristics were used, which needed be synchronized in an appropriate manner. Proper preparation of data for the calculation of dynamic coefficients of bearings is the most time-consuming aspect related to their determination. The "Signal" program was an effective tool for selecting signals, performing subtraction operations on them, and saving them in an appropriate form.

During the signal preparation process, the reference signal in the immediate vicinity of the excitation signal was subtracted from each signal registered by the eddy current sensor after the excitation. This procedure was performed as one operation for two bearings. For the entire speed range, it was relatively easy to find a reference signal, but around resonance speeds (4000 rpm) this task became more difficult as the signal shifted in phase. The frequency of sine wave signal changes measured by the eddy current sensor (e.g. in the *X* direction in one support) was different from the frequency of signal changes generated by the eddy current sensor placed in the same direction on the other support.

Dynamic coefficients of bearings were calculated in the "Calculation" program. The program input values were excitation force and response signals of the system, both previously prepared in the "Signal" program. For each of the 11 rotational speeds analyzed, 10 sets of stiffness, damping and mass coefficients were determined, and then 4 sets with the highest values of standard deviation were rejected. The mean value and standard deviation were calculated from the six sets of stiffness and damping coefficients for each speed.

Stiffness, damping and mass coefficients were determined simultaneously for two bearings. Bearing no. 1 was closer to the coupling connecting the shaft of the laboratory test rig with the drive motor shaft, and bearing no. 2 was located on the other side of the rotor. The excitation force was applied by means of an impact hammer near the disk located between bearing supports. The influence of the coupling on the results of the experimental tests is significant. In order to obtain correct results of calculation of dynamic coefficients of bearings, the influence should be minimized. A rigid coupling was used in experimental tests. The use of a coupling allowing for damping torsional vibrations and better compensation of the rotor axis displacement will have a positive effect on the results obtained. Due to the significant influence of the coupling on the calculation results, the bearing further away from the coupling, i.e. bearing no. 2, should be used for comparisons with numerical calculations.

The calculated stiffness, damping and mass coefficients over the entire speed range change in an expected way. The stiffness coefficients for the lowest tested speed of 2250 rpm are about 20 000 N/m. The highest stiffness of bearing no. 2 was observed in the *YX* direction. The stiffness coefficients in the main *XX* and *YY* directions are approximately twice as low. All values decrease smoothly up to 3750 rpm, followed by an increase when nearing resonance speed. The standard deviation values increased as the rotor speed increased, which means that the calculated values are less repeatable.

The highest values of damping coefficients in the second bearing were achieved for lower rotational speeds and decreased while speed increased. Their values start at approximately 110 N·s/m for the main coefficient c_{xx} and 65 N·s/m for the c_{yy} coefficient. The values of cross-coupled coefficients oscillate in the 20–30 N·s/m range. Initially all values were positive. Near resonance speeds the c_{xy} coefficients had negative values. At higher speeds (from approximately 5500 to 6000 rpm), the values of all damping coefficients (both main and cross-coupled) were approximately 30 N·s/m.

The mass coefficients calculated for the second bearing assume values ranging from approximately −0.4 kg for the *XX* and *XY* directions to 0.4 kg for the *YY* and *YX* directions. It is interesting to note that if, in the diagram illustrating mass coefficients, the calculated values using a straight line or a lower order polynomial curve are approximated, the values of the main coefficients change from positive values at lower speeds to the same negative values for the highest speeds tested. These curves pass through a zero value near the resonance speeds. This was in line with the fact that the main axis of the ellipse of vibrations changes its angle when passing through resonance speed.

9

Numerical Calculations of Bearing Dynamic Coefficients

This chapter presents a method of numerical determination of stiffness and damping coefficients of hydrodynamic radial bearings using methods with linear and non-linear calculation algorithms. The structure of numerical models prepared on the basis of previously conducted experimental tests is presented. The results obtained on the basis of linear and non-linear calculation models are included. At the end of the chapter, the numerical calculations are verified by comparing the displacements obtained from the calculated stiffness and damping coefficients with the results of the experimental tests.

Numerical calculations were performed with the use of unique linear and non-linear numerical models developed in the Institute of Fluid Flow Machinery at the Polish Academy of Sciences under the management of Professor Jan Kiciński.

9.1 Method of Calculating Dynamic Coefficients of Bearings

The procedure for numerical determination of the stiffness and damping coefficients of a lubricating film is quite complex. Programs from the MESWIR series were used for calculations (Breńkacz and Żywica 2016b). At the beginning, a method using a linear calculation algorithm shall be presented, which is adequate to solve the problem presented in Figure 1.2a. The next step shows how to extend it to carry out non-linear calculations. Only a brief description of equations, which were used for calculations, is presented in this monograph. The full development of equations is presented elsewhere (Kiciński 1994, 2006). The Reynolds equation for a bearing with a fixed bearing housing can be written using formula (9.1):

$$\frac{\partial}{\partial x}\left(\frac{\partial p}{\partial x}A\right)+\frac{\partial}{\partial z}\left(\frac{\partial p}{\partial z}A\right)=\left[\omega R\left(\frac{\partial h}{\partial x}-\frac{\partial B}{\partial x}\right)+\frac{\partial h}{\partial t}\right] \tag{9.1}$$

where A and B describe the relationships (9.2), while the coordinate system and the designations are in accordance with Figure 9.1.

Bearing Dynamic Coefficients in Rotordynamics: Computation Methods and Practical Applications,
First Edition. Łukasz Breńkacz.

Companion website: www.wiley.com/go/brenkacz/bearingdynamiccoefficients

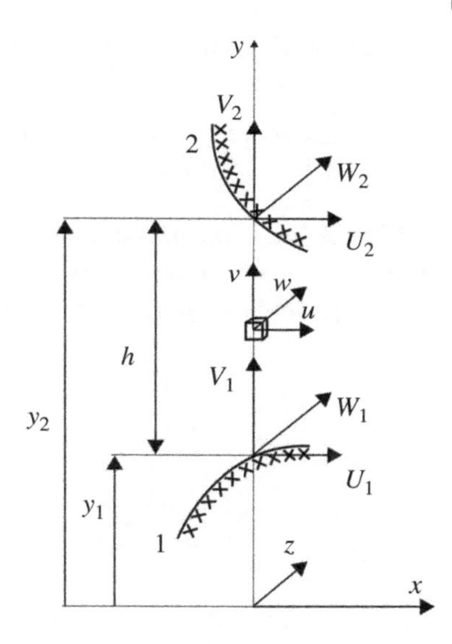

Figure 9.1 Coordinate system at a selected point in the lubrication gap. *Source:* Based on Kiciński (2005).

$$\left.\begin{aligned} A &= \int_0^h \left[\int_0^y \frac{y}{\mu}\mathrm{d}y - \frac{\int_0^h \frac{y}{\mu}\mathrm{d}y}{\int_0^h \frac{1}{\mu}\mathrm{d}y} \int_0^y \frac{1}{\mu}\mathrm{d}y \right] \mathrm{d}y \\ B &= \int_0^h \left[\frac{\int_0^y \frac{1}{\mu}\mathrm{d}y}{\int_0^h \frac{1}{\mu}\mathrm{d}y} \right] \mathrm{d}y \end{aligned}\right\} \tag{9.2}$$

For further consideration, it is convenient to assume the dimensionless form of the equation by introducing the following relationships, typical of radial plain bearings (9.3):

$$\psi = \frac{x}{R}, \quad Y = \frac{y}{h}, \quad H = \frac{h}{\Delta R}, \quad Z = \frac{2x}{L}, \quad M = \frac{\mu}{\mu_0}, \quad \tau = \Omega t \tag{9.3}$$

where ΔR is the radial bearing backlash, L is the bearing housing width, R is the journal radius, h is the lubrication gap height (between surfaces 1 and 2), μ_0 is the dynamic viscosity of oil at temperature T_0, Ω is the journal angular velocity, and t is time. The value of the hydrodynamic pressure p can then be expressed in dimensionless form as (9.4) (where d is journal diameter), while the expressions A and B of (9.2) can be written in dimensionless form as Eq. (9.5):

$$= \frac{p\left(\frac{\Delta R}{R}\right)^2}{Ld\mu_0{}^{\Omega}} \tag{9.4}$$

$$\overline{A} = A\frac{\mu_0}{h^3} \text{and}\, \overline{B} = B\frac{1}{h} \tag{9.5}$$

Equation (9.1) can now be written in dimensionless form as (9.6):

$$\frac{\partial}{\partial \psi}\left(H^3\overline{A}\frac{\partial \Pi}{\partial \psi}\right)+\left(\frac{d}{L}\right)^2\frac{\partial}{\partial Z}\left(H^3\overline{A}\frac{\partial \Pi}{\partial Z}\right)=\left[\frac{\partial H}{\partial \psi}\left(\overline{B}-1\right)+H\frac{\partial \overline{B}}{\partial \psi}-\frac{\partial H}{\partial \tau}\right] \tag{9.6}$$

After several transformations, taking into account the components of the lubricating film reaction and components of displacement, formulae for dimensionless stiffness coefficients $\gamma_{i,k}$ and damping $\beta_{i,k}$ (9.7) can be obtained:

$$\gamma_{i,k}=k_{i,k}\frac{\left(\Delta R/R\right)^2\Delta R}{dL\mu_0\omega}=k_{i,k}S_0\frac{\Delta R}{P_{st}}$$

$$\beta_{i,k}=c_{i,k}\frac{\left(\Delta R/R\right)^2\Delta R}{dL\mu_0}=c_{i,k}\Omega S_0\frac{\Delta R}{P_{st}} \tag{9.7}$$

where S_0 is the Sommerfeld number and W_{x0} and W_{y0} are the components of the film reaction at the point of static equilibrium $(O_c)_{st}$, according to Eq. (9.8):

$$S_0=\frac{P_{st}\left(\Delta R/R\right)^2}{dL\mu_0},\quad P_{st}=\sqrt{W_{x0}{}^2+W_{y0}{}^2} \tag{9.8}$$

The traditional approach to determining these coefficients is to take small increments and calculate the reactions for their subsequent values, which means solving the Reynolds equation four times and then calculating the derivatives. This method is not very accurate as the result depends on the size of the assumed increments.

At this point it is possible to introduce a very accurate and fast method of calculating stiffness and damping coefficients based on the perturbation. The starting point for such an analysis is the solution of the Reynolds equation (9.6). The perturbation method consists in developing into Taylor series at the point of static equilibrium $(O_c)_{st}$ of all the components of Eq. (9.6) and reflecting the mutual relations between forces and displacements (in this case between the pressure Π and the shape of lubrication gap H) according to the disturbing parameters X_c, Y_c, $\tilde{X}_c$, and $\tilde{Y}_c$.

The shape of the lubrication gap H is a function of both surface coordinates. There is no need to develop integrals into the Taylor series, because they are functions of viscosity, and in the adopted model viscosity does not depend directly on the pressure Π. In practice, the viscosity M depends directly on the pressure Π, but this dependency is minimal for small relative displacements of the journal and the bearing housing – hence it can be neglected. Development into the Taylor series of expressions for the pressure Π and the shape of the gap H takes the form (9.9):

$$\left.\begin{aligned}\Pi&=\Pi_{st}+\left(\frac{\partial \Pi}{\partial X_c}\right)_{st}X_c+\left(\frac{\partial \Pi}{\partial Y_c}\right)_{st}Y_c+\left(\frac{\partial \Pi}{\partial \widetilde{X}_c}\right)_{st}\widetilde{X}_c+\left(\frac{\partial \Pi}{\partial \widetilde{Y}_c}\right)_{st}\widetilde{Y}_c\\H&=H_{st}+\left(\frac{\partial H}{\partial X_c}\right)_{st}X_c+\left(\frac{\partial H}{\partial Y_c}\right)_{st}Y_c\end{aligned}\right\}. \tag{9.9}$$

The searched stiffness and damping coefficients can be presented as (9.10):

$$
\left.
\begin{aligned}
\gamma_{xx} &= -\frac{1}{4}\int_{-1}^{1}\int_{\psi_1}^{\psi_2}\left(\frac{\partial \Pi}{\partial X_c}\right)_{st}\cos\psi\, \mathrm{d}\psi \mathrm{d}Z \\
\gamma_{yx} &= -\frac{1}{4}\int_{-1}^{1}\int_{\psi_1}^{\psi_2}\left(\frac{\partial \Pi}{\partial X_c}\right)_{st}\sin\psi\, \mathrm{d}\psi \mathrm{d}Z \\
\gamma_{xy} &= -\frac{1}{4}\int_{-1}^{1}\int_{\psi_1}^{\psi_2}\left(\frac{\partial \Pi}{\partial Y_c}\right)_{st}\cos\psi\, \mathrm{d}\psi \mathrm{d}Z \\
\gamma_{yy} &= -\frac{1}{4}\int_{-1}^{1}\int_{\psi_1}^{\psi_2}\left(\frac{\partial \Pi}{\partial Y_c}\right)_{st}\sin\psi\, \mathrm{d}\psi \mathrm{d}Z \\
\beta_{xx} &= -\frac{1}{4}\int_{-1}^{1}\int_{\psi_1}^{\psi_2}\left(\frac{\partial \Pi}{\partial \widetilde{X}_c}\right)_{st}\cos\psi\, \mathrm{d}\psi \mathrm{d}Z \\
\beta_{yx} &= -\frac{1}{4}\int_{-1}^{1}\int_{\psi_1}^{\psi_2}\left(\frac{\partial \Pi}{\partial \widetilde{X}_c}\right)_{st}\sin\psi\, \mathrm{d}\psi \mathrm{d}Z \\
\beta_{xy} &= -\frac{1}{4}\int_{-1}^{1}\int_{\psi_1}^{\psi_2}\left(\frac{\partial \Pi}{\partial \widetilde{Y}_c}\right)_{st}\cos\psi\, \mathrm{d}\psi \mathrm{d}Z \\
\beta_{yy} &= -\frac{1}{4}\int_{-1}^{1}\int_{\psi_1}^{\psi_2}\left(\frac{\partial \Pi}{\partial \widetilde{Y}_c}\right)_{st}\sin\psi\, \mathrm{d}\psi \mathrm{d}Z
\end{aligned}
\right\}
\tag{9.10}
$$

In order to determine the distributions $(\partial\Pi/\partial X_c)_{st}$ and $(\partial\Pi/\partial\tilde{Y}_c)_{st}$, it is necessary to develop the pressure derivatives Π and gap shape H included in Eq. (9.6) into a Taylor's series as well. By inserting these values into the Reynolds equation (9.6) and grouping the expressions accordingly, we obtain a system of equations:

$$
\left.
\begin{aligned}
\frac{\partial}{\partial\psi}\left[H_{st}{}^{3}\bar{A}\frac{\partial}{\partial\psi}\left(\frac{\partial\Pi}{\partial X_c}\right)_{st}\right]+\left(\frac{d}{L}\right)^{2}\frac{\partial}{\partial Z}\left(H_{st}{}^{3}\bar{A}\frac{\partial}{\partial Z}\left(\frac{\partial\Pi}{\partial X_c}\right)_{st}\right) &= R_x \\
\frac{\partial}{\partial\psi}\left[H_{st}{}^{3}\bar{A}\frac{\partial}{\partial\psi}\left(\frac{\partial\Pi}{\partial \widetilde{X}_c}\right)_{st}\right]+\left(\frac{d}{L}\right)^{2}\frac{\partial}{\partial Z}\left(H_{st}{}^{3}\bar{A}\frac{\partial}{\partial Z}\left(\frac{\partial\Pi}{\partial \widetilde{X}_c}\right)_{st}\right) &= R_{\tilde{x}} \\
\frac{\partial}{\partial\psi}\left[H_{st}{}^{3}\bar{A}\frac{\partial}{\partial\psi}\left(\frac{\partial\Pi}{\partial Y_c}\right)_{st}\right]+\left(\frac{d}{L}\right)^{2}\frac{\partial}{\partial Z}\left(H_{st}{}^{3}\bar{A}\frac{\partial}{\partial Z}\left(\frac{\partial\Pi}{\partial Y_c}\right)_{st}\right) &= R_y \\
\frac{\partial}{\partial\psi}\left[H_{st}{}^{3}\bar{A}\frac{\partial}{\partial\psi}\left(\frac{\partial\Pi}{\partial \widetilde{Y}_c}\right)_{st}\right]+\left(\frac{d}{L}\right)^{2}\frac{\partial}{\partial Z}\left(H_{st}{}^{3}\bar{A}\frac{\partial}{\partial Z}\left(\frac{\partial\Pi}{\partial \widetilde{Y}_c}\right)_{st}\right) &= R_{\tilde{y}} \\
\frac{\partial}{\partial\psi}\left[H_{st}{}^{3}\bar{A}\left(\frac{\partial\Pi}{\partial\psi}\right)_{st}\right]+\left(\frac{d}{L}\right)^{2}\frac{\partial}{\partial Z}\left(H_{st}{}^{3}\bar{A}\left(\frac{\partial\Pi}{\partial Z}\right)_{st}\right) &= R_{st}
\end{aligned}
\right\}.
\tag{9.11}
$$

One can notice that the fifth "undisturbed" equation from the system of Eq. (9.11)) is simply the Reynolds equation (9.6) at the point of static equilibrium, i.e. $\partial H/\partial\tau = 0$. Also, the left side of the disturbing differential equations is similar to the Reynolds equation, as proven by substitution of $\partial\Pi/\partial X_c$ with the pressure Π. This means that all equations from the system (9.11) can be solved according to an identical numerical scheme. This fact is important for the convenient form as well as for the speed of numerical calculations. Taking into account the appropriate geometric relationships, the right side of Eq. (9.11) can be written as (9.12):

$$\left.\begin{array}{l}
R_x = \dfrac{\partial R_{st}}{\partial X_c} + R_{xA} \quad R_y = \dfrac{\partial R_{st}}{\partial Y_c} + R_{yA} \\
R_{\dot{x}} = \cos\psi \quad R_{\dot{y}} = \sin\psi \\
R_{st} = \left(\dfrac{\partial H}{\partial \psi}\right)_{st}\left(\overline{B}-1\right) + H_{st}\dfrac{\partial \overline{B}}{\partial \psi} \\
\dfrac{\partial R_{st}}{\partial X_c} = \left(\overline{B}-1\right)\sin\psi - \dfrac{\partial \overline{B}}{\partial \psi}\cos\psi \\
\dfrac{\partial R_{st}}{\partial Y_c} = \left(\overline{B}-1\right)\cos\psi - \dfrac{\partial \overline{B}}{\partial \psi}\sin\psi \\
R_{xA} = 3\left\{\dfrac{R_{st}}{H_{st}}\cos\psi - H_{st}\overline{A}\left(\dfrac{\partial \Pi}{\partial \psi}\right)_{st}\left[H_{st}\sin\psi + \left(\dfrac{\partial H}{\partial \psi}\right)_{st}\cos\psi\right]\right. \\
\qquad \left. -\left(\dfrac{d0}{L}\right)^2 H_{st}\overline{A}\left(\dfrac{\partial \Pi}{\partial Z}\right)_{st}\left(\dfrac{\partial \Pi}{\partial Z}\right)_{st}\cos\psi\right\} \\
R_{yA} = 3\left\{\dfrac{R_{st}}{H_{st}}\sin\psi - H_{st}\overline{A}\left(\dfrac{\partial \Pi}{\partial \psi}\right)_{st}\left[H_{st}\cos\psi + \left(\dfrac{\partial H}{\partial \psi}\right)_{st}\sin\psi\right]\right. \\
\qquad \left. -\left(\dfrac{d}{L}\right)^2 H_{st}\overline{A}\left(\dfrac{\partial \Pi}{\partial Z}\right)_{st}\left(\dfrac{\partial \Pi}{\partial Z}\right)_{st}\sin\psi\right\}
\end{array}\right\}. \tag{9.12}$$

Dimensionless stiffness and damping coefficients of the oil film defined by Eq. (9.10) can be determined by solving the disturbing differential Eq. (9.11) taking into account their right sides in the form of (9.12). However, using the relationships in (9.7) it is easy to obtain the values of these coefficients in dimensional form.

A non-linear description is necessary when there are large journal displacements within the lubrication gap (Fertis 2010), i.e. in the situation shown in Figure 1.2b, when Eq. (1.9) is true. In this case the perturbation method can also be used, but it must be modified accordingly. It should be assumed that what has so far been marked was the static equilibrium point of the journal $(O_c)_{st}$, now indicates the position of the journal O_c for the selected moment of time t_k with its large displacements, as shown in Figure 1.2b. For a moment of time t_k specified in such a way, the Reynolds equation (9.6) with its dynamic element $\partial H/\partial\tau$, i.e. derivatives $\partial\varepsilon/\partial t$ and $\partial\gamma/\partial t$, is solved.

If for sufficiently small intervals $\Delta t = t_k - t_{k-1}$ we are able to find sufficiently small intervals Δx and Δy in which it can be assumed that oil film properties are constant and described by "dynamic" stiffness and damping coefficients, which are momentary and constant in these intervals, then the whole procedure for determining these coefficients using the perturbation method can also be applied in the case of a non-linear description. In the expressions defining the right sides of the disturbing equations in (9.12) it is enough to place a new value on R_{st}, which will now be the right side of the Reynolds equation (9.6) with a dynamic component, i.e. (9.13):

$$R_{st} = R_{st}^{*} = \frac{\partial H}{\partial \psi}\left(\bar{B}-1\right)+\frac{\partial \bar{B}}{\partial \psi}-\frac{\partial H}{\partial \tau}. \tag{9.13}$$

It is also necessary in the disturbing equation (9.11) instead of the Π pressure calculated at the point of static equilibrium, to provide Π^* as a result of the solution of the "dynamic" Reynolds equation (9.6) for a given moment of time t_k, i.e. (9.14):

$$\Pi = \Pi^{*} \tag{9.14}$$

Taking into account the relationships (9.13) and (9.14), the entire perturbation procedure for determining the stiffness and damping coefficients of the lubricating film can be repeated as in the case of the linear description, i.e. the disturbing equation (9.11) can be solved taking into account the relationships in (9.12) and finally the relationships in (9.10).

In order to accurately calculate the dynamic coefficients of bearings, in addition to the Reynolds equation, it is also necessary to solve the energy equation, the conductivity equation, the lubrication gap shape equation, the interfering equations and, finally, the equations defining the stiffness and damping coefficients of the lubricating film. They form a coherent logical whole, so they are linked to each other. This fact causes difficulties in solving them. Selection of an appropriate algorithm is a matter of many years of experience and trials of research teams dealing with this problem (Kiciński 1994). In the MESWIR series of programs, the Institute of Fluid Flow Machinery at the Polish Academy of Sciences adopted an algorithm, which has proved its worth in many practical applications. A block diagram with the most important steps of the algorithm is shown in Figure 9.2. The energy and conductivity equations are connected by a separate iterative loop. This is due to the fact that the boundary conditions defined in the energy equation for the conductivity equation (temperature gradients) have a very large impact on the stability of the numerical solution. The use of a separate iterative loop and the use of additional relaxation methods with a relaxation coefficient value much lower than 1 allows for stable solutions.

9.2 Calculation of Dynamic Coefficients of Bearings Using a Method with Linear Calculation Algorithm

In the first stage of numerical calculations, the stiffness and damping coefficients were calculated for a linear isothermal model of the bearing in the KINWIR program (part of the MESWIR package). The kinetostatic reactions of supports, the kinetostatic journal position and the stiffness and damping coefficients of the lubricating film for the position of kinetostatic equilibrium are calculated in this way. These data were used in the LDW program, which was used as a means to determine the vibration trajectory of bearing journals.

The numerical model was built in such a way to reflect the laboratory test rig presented in Chapter 3 as accurately as possible. It consists of Timoshenko-type beam elements

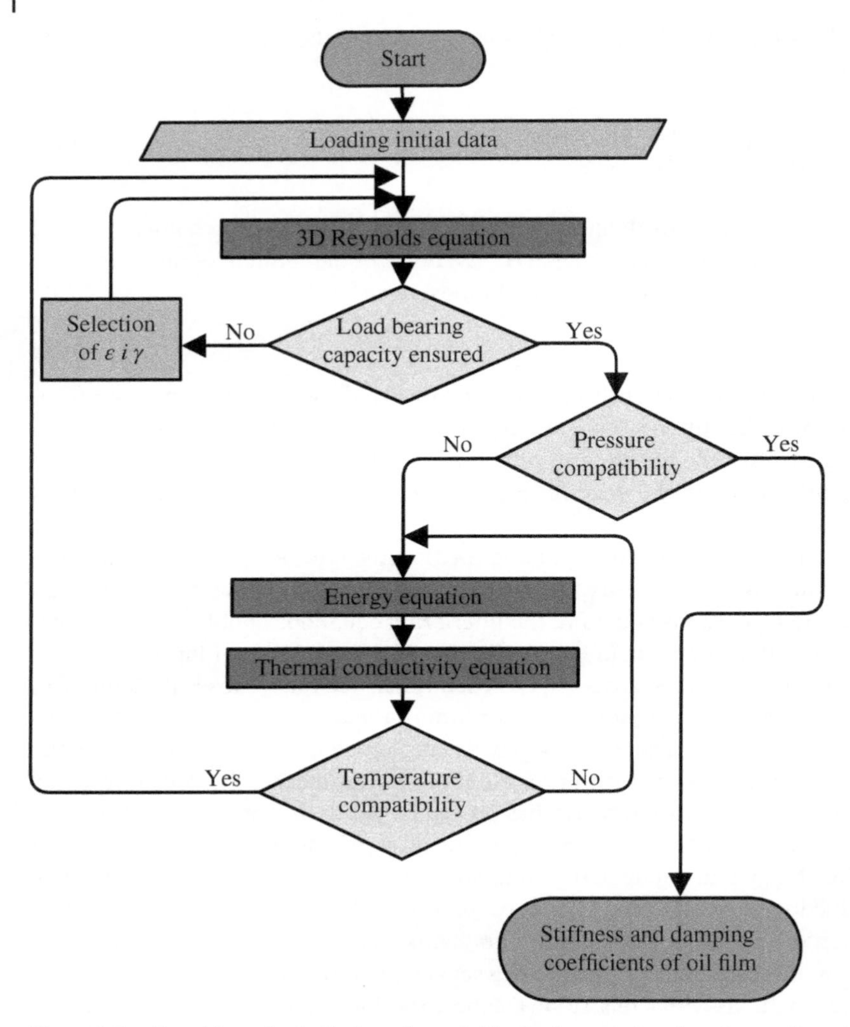

Figure 9.2 Algorithm of calculation of a radial hydrodynamic bearing. *Source:* Based on Kiciński (2005).

(Figure 9.3). Two hydrodynamic bearings were placed in nodes 7 and 27. In node 17 of the rotor a concentrated force was located, which in numerical calculations symbolizes the disk. The value of this force was assumed to be the product of the mass of the disk and gravity. Parameters of the numerical model are listed in Table 9.1.

The stiffness of bearing supports is assumed to be $-0.2 \cdot 10^{15}$ N/m. On the basis of previous analyses, it can be concluded that this value corresponds to the rigid support of the rotor. In the numerical model prepared in such a way, only the properties of the lubricating film are taken into account for further analysis of the rotor dynamics, rather than the support structure. The damping of supports is assumed to be 30 000 N·s/m, while the mass of one support is 1 kg. Due to the very high stiffness of the foundation, the mass and damping values did not affect the dynamic properties of the rotor.

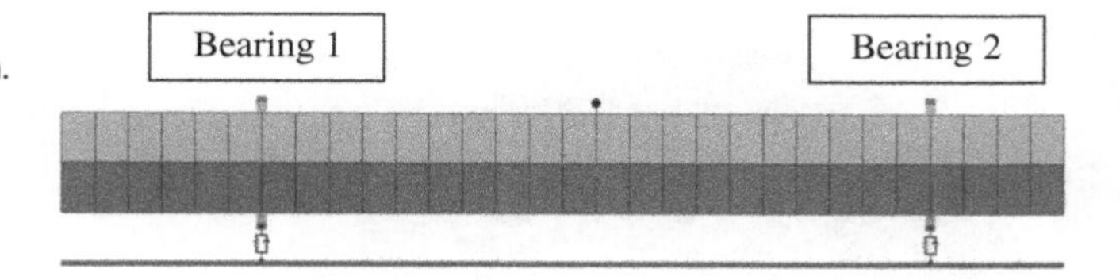

Figure 9.3 Numerical rotor model in the KINWIR program.

Table 9.1 Parameters of the numerical model in the KINWIR program.

Parameter	Value
Young's modulus	210 GPa
Rotor material density	7860 kg/m^3
Poisson's ratio	0.3
External damping coefficient (alpha)	1.2
Internal damping coefficient (beta)	$0.8 \cdot 10^{-4}$
Number of beam elements	30
Length of the beam element	30 mm
Diameter of the beam element in regards to mass	19.02 mm
Diameter of the beam element in regards to inertia	19.02 mm

A numerical model of a hydrodynamic radial bearing is shown in Figure 9.4. The bearing housing width is 12.6 mm. The absolute radial clearance is 76 μm. The number of elements of the numerical grid in the axial direction of the bearing is 4. The height of the lubrication gap is 0.84 mm. The oil flowed from both sides of the bearing simultaneously. Differences can be observed between the numerical and real hydrodynamic bearing model because in experimental tests oil was supplied by means of a cylindrical hole. Simplification in the numerical model is necessary for calculations, however, the lubrication gap surface is the same as in the real bearing. The oil supply pressure was the same as in the experimental tests and equaled 160 000 Pa. The pressure on the edge of the bearing housing was equal to atmospheric pressure. ISO 13 oil was used as a lubricant in hydrodynamic bearings. Its viscosity was 0.01105 N·s/m^2 at 30 °C. In the numerical model two identical bearings were modeled. The bearing housings are arranged in a single line, without misalignment.

The initial position of the rotor journal in the bearing is described by an eccentricity (ε – distance from the journal center to the bearing housing center) of 0.3. The angle of the line defined by the center of the bearing housing and the center of the journal from the X axis in counterclockwise direction (γ) is assumed to be 335°. The minimum size of the lubrication gap in numerical calculations was assumed to be 1 μm. The maximum permissible relative error of pressure distribution (accuracy of the Reynolds equation solution) was 0.001. The relative initial eccentricity increment was equal to 0.01. The admissible relative error in selecting the load capacity to the set external load was 0.01. The maximum absolute error summed with the relative error in the load-bearing capacity selection loop is

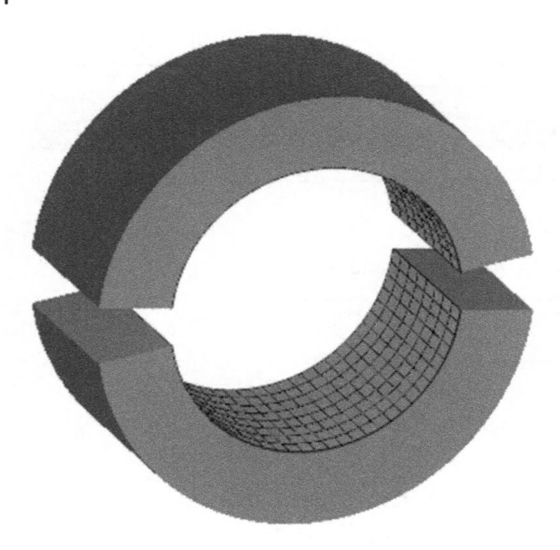

Figure 9.4 Numerical model of hydrodynamic bearing.

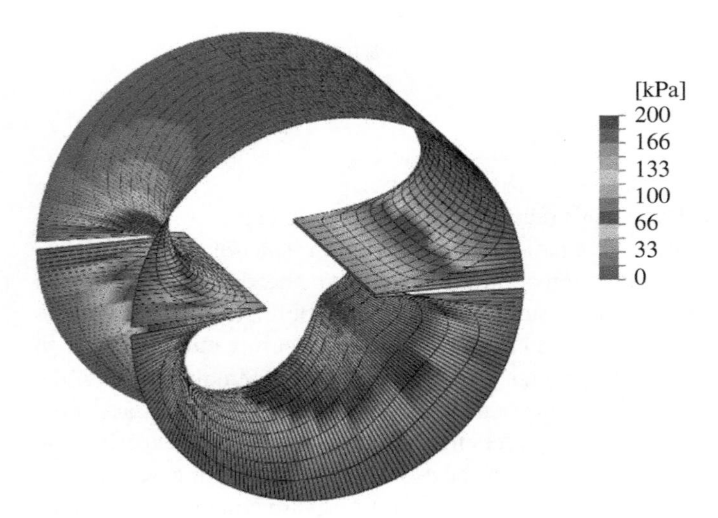

Figure 9.5 Presentation of the pressure distribution calculated for the static equilibrium point at a speed of 3250 rpm in the Kinwir program.

0.1 N. The maximum number of iterations in the pressure selection loop was 500 and in the load-bearing capacity selection loop was 2000. The maximum number of iterations in the EPS and GAM selection loop is 250. The relaxation coefficient in the GAM angle selection loop was equal to 0.1.

The basis for calculating the stiffness and damping coefficients of hydrodynamic radial bearings is the pressure distribution on the bearing housing. Figure 9.5 shows the pressure distribution calculated for a speed of 3250 rpm in the KINWIR program. To calculate the pressure distribution, the numerical grid was thickened. In the axial direction there are

Table 9.2 Stiffness and damping coefficients obtained from linear calculations in the KINWIR.

Rotational speed (rpm)	k_{xx}	k_{yy}	k_{xy}	k_{yx}	c_{xx}	c_{yy}	c_{xy}	c_{yx}
	(N/m)				(N·s/m)			
2000	447 000	407 000	−1430	−597 000	4020	3190	−2610	−2600
2500	401 000	382 000	18 900	−572 000	3020	2510	−1880	−1880
2750	387 000	366 000	27 400	−552 000	2630	2250	−1620	−1620
3000	369 000	354 000	38 700	−534 000	2310	2060	−1420	−1430
3250	363 000	350 000	48 700	−531 000	2110	1940	−1300	−1300
3500	360 000	340 000	53 100	−523 000	1920	1790	−1160	−1160
3750	338 000	329 000	66 900	−507 000	1730	1690	−1050	−1050
4000	331 000	323 000	73 500	−499 000	1590	1580	−957	−958
4250	335 000	320 000	78 700	−503 000	1510	1510	−897	−897
4500	324 000	314 000	88 800	−497 000	1400	1450	−833	−834
4750	326 000	311 000	93 700	−499 000	1330	1390	−782	−783
5000	309 000	301 000	102 000	−492 000	1250	1310	−712	−713
5250	307 000	302 000	109 000	−494 000	1200	1270	−674	−675
5500	306 000	299 000	115 000	−495 000	1140	1230	−639	−640
5750	304 000	296 000	122 000	−497 000	1100	1200	−609	−610
6000	277 000	296 000	137 000	−507 000	1090	1170	−564	−564

now 16 divisions, while the bearing has been divided into 256 parts along the perimeter. The figure shows the classic position of the lubricating film. The maximum pressure in the bearing housing is 160 kPa.

The numerical model of the rotor, similarly to the one used in experimental tests, is asymmetrical. On the one hand, it is 60 mm longer due to the presence of the coupling. This makes the calculated values for stiffness and damping coefficients differ for the two bearings, however, these differences are small. The results of stiffness and damping coefficients calculated for the bearing in Figure 9.3 shown as bearing 2 (located at the free end of the shaft) were used to compare the results of various calculation methods in further work. The results of stiffness and damping coefficients for bearing no. 2, for rotational speeds from 2000 to 6000 rpm, are presented in Table 9.2.

Figure 9.6 shows the stiffness coefficients in the main and cross-coupled directions, calculated on the basis of the numerical model described above. The coefficients in the main directions (*XX* and *YY*) are similar and change from approximately 400 000 N/m (at 2000 rpm) to approximately 300 000 N/m (at 6000 rpm). The cross-coupled coefficients in the *XY* direction at 2000 rpm are close to zero and increase as the rotational speed increases, reaching a value of approximately 150 000 N/m at a rotor speed of 6000 rpm. The values of cross-coupled stiffness coefficients in the *YX* direction are negative. Their absolute values are greater than those of the main stiffness coefficients and decrease as the rotational speed increases from −600 000 to approximately −500 000 N/m at 4000 rpm. They remain constant at higher speeds.

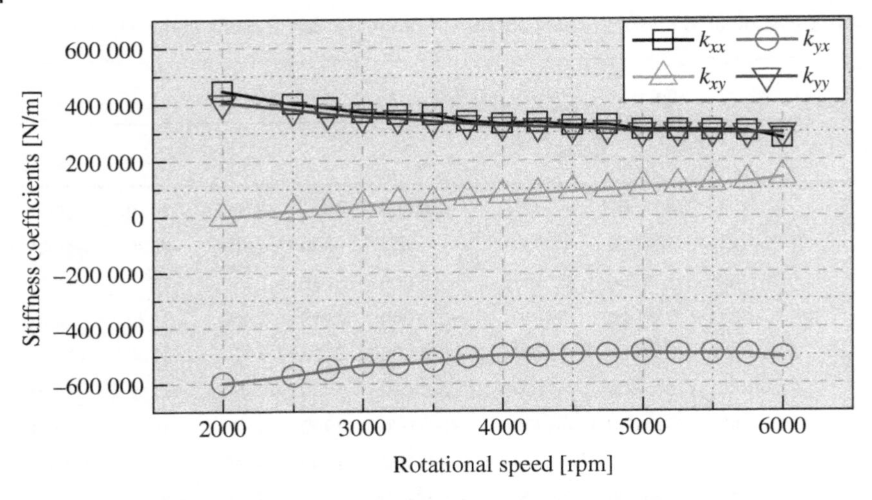

Figure 9.6 Bearing no. 2 stiffness coefficients calculated using a linear algorithm in the KINWIR program.

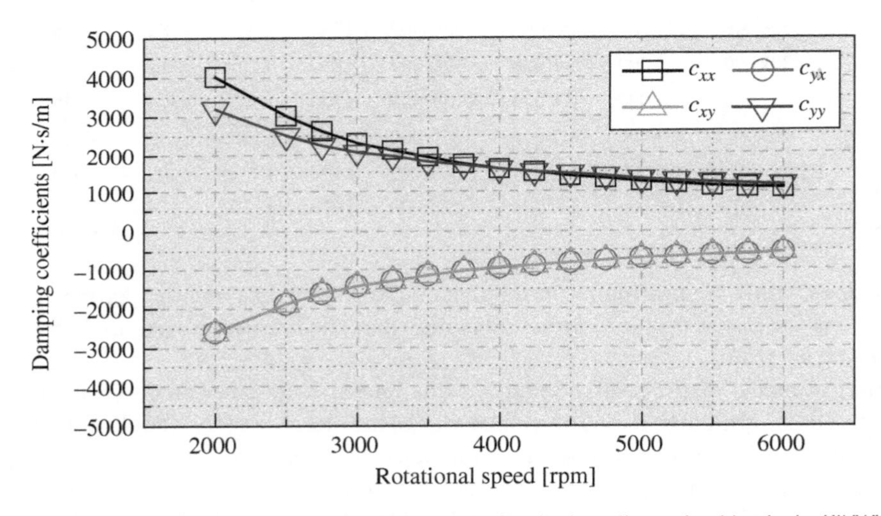

Figure 9.7 Bearing no. 2 damping coefficients calculated using a linear algorithm in the KINWIR program.

Figure 9.7 shows the calculation results of the damping coefficients. For a speed of 2000 rpm their values are 4000 N·s/m in the *XX* direction, about 3000 N·s/m in the *YY* direction and −2500 N·s/m in cross-coupled directions. As the rotational speed increases, their absolute values decrease and at 6000 rpm they are approximately 1000 N·s/m for the main directions and approximately −500 N·s/m for cross-coupled directions. At lower speeds, the coefficients for the *XX* direction are slightly higher than those calculated in the *YY* direction, but at a speed of approximately 4000 rpm this situation changes and slightly higher damping values can be observed for the damping coefficient in the *YY* direction. The values of cross-coupled damping coefficients in the *XY* and *YX* directions are the same for the entire speed range (Kiciński 1994). This relationship shall apply to all numerical calculation results presented in this work.

9.3 Calculation of Dynamic Coefficients of Bearings Using a Method with Non-linear Calculation Algorithm

Non-linear calculations were performed in the NLDW program. The numerical model in the NLDW program is similar to the one presented for linear calculations in Section 9.2, therefore only the differences between the models will be described.

A numerical view of the rotor is shown in Figure 9.8. The KINWIR program uses concentrated force as a model of the disk, in NLDW there is an element called "disk." It is placed on node 17 of the rotor (between bearings). The outer diameter of the disk was 152.4 mm and its inner diameter was 19.02 mm. The thickness of the disk was 20 mm. Aluminum with a density of 2720 kg/m^3 was used as the disk material.

In numerical non-linear calculation models, calculations are performed for each rotor position. Therefore, there is no single pressure field distribution for constant speed, as is the case for linear calculations. It changes at any moment in time. Figure 9.9 shows the results of calculations for a speed of 3250 rpm for 4 rotor positions calculated every 90°. As the rotor journal position changes, the distribution and maximum pressure generated in the hydrodynamic bearing changes. The maximum pressure value of 350 kPa is shown in the lubrication wedge shown in Figure 9.9d. In the other phases of operation, the pressure did not exceed 160 kPa.

On the basis of non-linear calculations, it is not a single value for stiffness and damping coefficients that is obtained, but a set of values that change over time. A list of minimum and maximum values of stiffness and damping coefficients is presented in Tables 9.3 and 9.4, respectively. The differences between the minimum and maximum values are very large. When operating at constant speed, the values of the main coefficients are often negative and the sign changes.

The graphical representation of changes in dynamic coefficients of bearings is a better illustration of the changes occurring during operation at constant speed. Figures 9.10–9.15 show the values of stiffness and damping coefficients calculated for three speeds in the NLDW program (3250, 3750, and 5500 rpm). They illustrate exemplary results of calculations made for speeds lower than the resonance speed for resonance and ultra-resonance

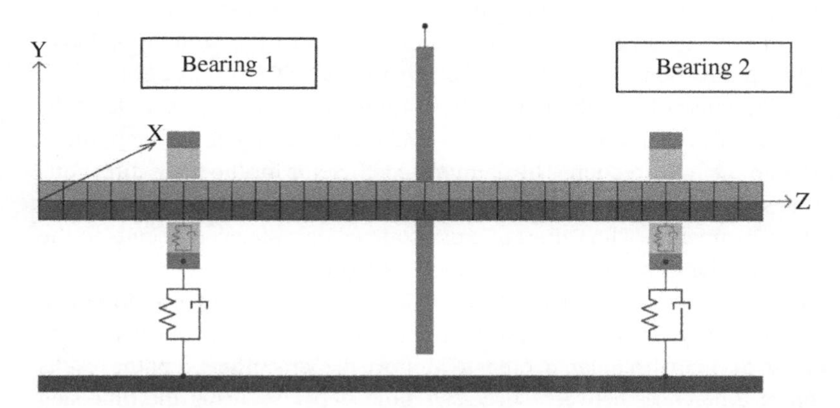

Figure 9.8 Numerical model of the rotor, bearing and disk in the NLDW program.

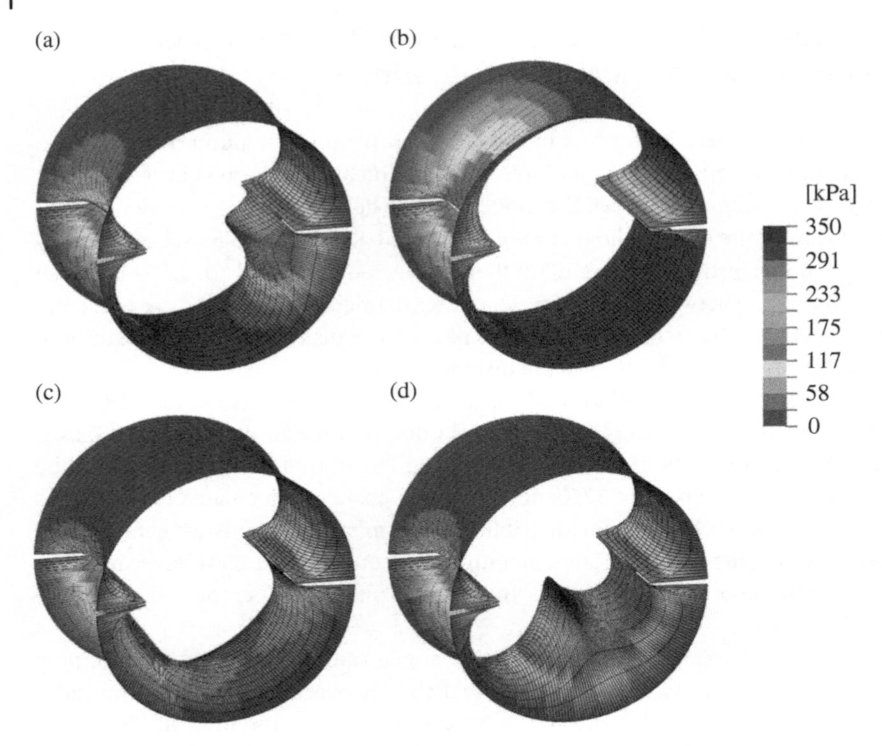

Figure 9.9 Pressure distribution calculated using the NLDW program for (a–d) 4 journal positions (in 90° increments) at 3250 rpm.

speed. Due to the high variability of the calculated values of coefficients, on each of the diagrams the courses for 3 rotor shaft revolutions are presented. As in the case of linear calculations, the results of dynamic coefficients of bearings calculated for the second bearing (located further away from the coupling) are presented.

For a speed of 3250 rpm the changes in stiffness and damping coefficients occurring periodically for each rotation of the rotor are observed. The main stiffness coefficients in the *XX* direction change from −393 680 to 3 061 800 N/m, and in the *YY* direction from −602 730 to 2 045 300 N/m. In the *XY* direction changes from −560 420 to 1 923 400 N/m can be observed. These are similar oscillations as for coefficients in the *YY* direction. The cross-coupled stiffness coefficient in the *YX* direction changes from −3 424 700 to 90 049 N/m. Amplitudes were similar to those in the *XX* direction. The damping coefficients for the main directions change from 199 N·s/m to about 12 993 N·s/m for the *XX* direction and from zero to 11 394 N·s/m for the *YY* direction. The values of cross-coupled damping coefficients in the *XY* and *YX* directions are the same and change from −7588 to 1399 N·s/m.

For resonance speed in numerical calculations (i.e. at 3750 rpm), the courses of stiffness and damping coefficients have very high amplitudes, but they also change synchronously with the rotor speed. Their course is more irregular. In order to improve the "smoothness" of the results obtained (reduction of differences between successive time steps), reducing the time step could prove to be helpful. The stiffness coefficients change from −1 473 100 to 7 843 500 N/m.

Table 9.3 Minimum and maximum values of stiffness coefficients (N/m) calculated in the NLDW program.

Speed (rpm)	k_{xx}^{min}	k_{xx}^{max}	k_{yy}^{min}	k_{yy}^{max}	k_{xy}^{min}	k_{xy}^{max}	k_{yx}^{min}	k_{yx}^{max}
2000	−27 324	934 220	174 810	661 850	−57 450	166 870	−998 070	−203 360
2500	−152 430	1 150 700	−64 555	954 250	−114 610	531 010	−1 369 100	−5108
2750	−188 720	1 338 900	−132 980	1 081 200	−168 380	717 430	−1 607 900	86 309
3000	−203 540	1 782 500	−228 470	1 326 500	−265 700	1 050 100	−2 087 900	98 416
3250	−393 680	3 061 800	−602 730	2 045 300	−560 420	1 923 400	−3 424 700	90 049
3500	−990 310	3 771 300	130 840	3 949 500	−1 143 400	2 087 000	−4 509 800	393 830
3750	−1 473 100	7 843 500	20 454	7 647 600	−3 188 300	6 324 700	−6 864 900	2 438 100
4000	−252 660	4 386 500	−604 140	3 054 200	−697 700	1 221 000	−4 972 800	275 250
4250	−536 020	2 579 900	−516 400	2 098 600	−388 360	1 852 800	−3 184 100	72 633
4500	−297 840	2 166 000	−369 100	1 836 400	−312 520	1 473 300	−2 801 500	95 681
4750	−240 020	1 863 600	−309 090	1 644 500	−295 720	1 200 900	−2 524 000	100 300
5000	−226 400	1 648 400	−269 940	1 544 600	−273 040	1 053 700	−2 340 600	74 170
5250	−241 340	1 543 400	−288 170	1 508 000	−184 960	969 230	−2 429 800	55 714
5500	−268 840	1 478 900	−233 570	1 537 200	−22 943	956 560	−2 520 300	43 052
5750	−596 290	1 506 000	−146 170	1 597 000	34 756	1 099 500	−2 668 000	22 272

Table 9.4 Minimum and maximum values of damping coefficients (N·s/m) calculated in the NLDW program.

Speed (rpm)	c_{xx}^{min}	c_{xx}^{max}	c_{yy}^{min}	c_{yy}^{max}	c_{xy}^{min}	c_{xy}^{max}
2000	1672	6626	1953	5149	−3859	−1169
2500	416	7060	996	6064	−3969	229
2750	280	7448	500	6519	−4115	363
3000	212	8781	462	7629	−4885	629
3250	199	12 993	658	11 394	−7588	1399
3500	436	9307	737	15 468	−5154	3168
3750	572	17 047	725	13 235	−9243	7132
4000	335	11 565	777	14 294	−9025	2198
4250	221	9229	681	8807	−5365	1128
4500	202	7481	654	7363	−4271	897
4750	198	6259	538	6407	−3538	773
5000	209	5374	407	5825	−3058	718
5250	332	5107	594	6188	−3201	859
5500	453	4955	571	6549	−3391	991
5750	−5047	−588	−7046	−560	−1400	3639

The stiffness coefficients in the *YY* direction change from 20 454 to 7 647 600 N/m. Cross-coupled stiffness coefficients in the *XY* direction change from −3 188 300 to 6 324 700 N/m, whereas in the *YX* direction they change from −6 864 900 to 2 438 100 N/m. The damping coefficients for the main *XX* direction change from 572 N·s/m to about 17 047 N·s/m, whereas for the *YY* direction they change from 725 to 13 235 N·s/m. Cross-coupled damping coefficients in *XY* and *YX* directions change from −9243 to 7132 N·s/m.

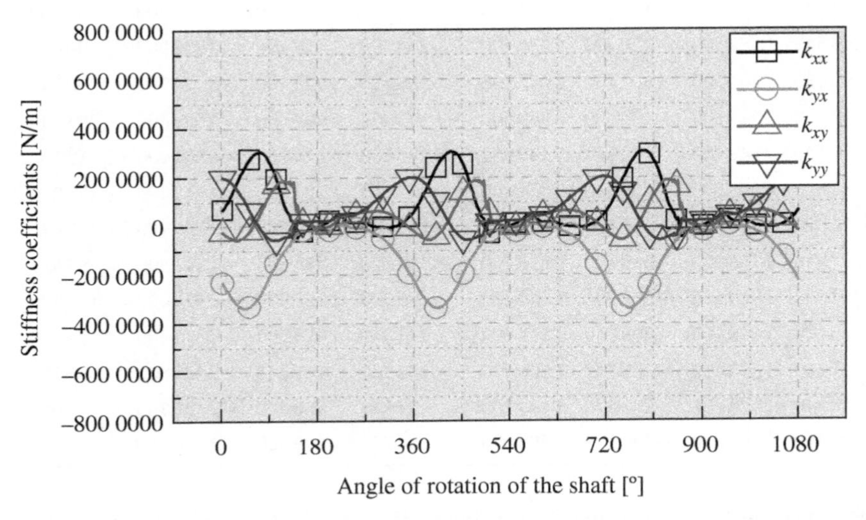

Figure 9.10 Stiffness coefficients calculated in the NLDW program for 3 rotor revolutions at a rotational speed of 3250 rpm.

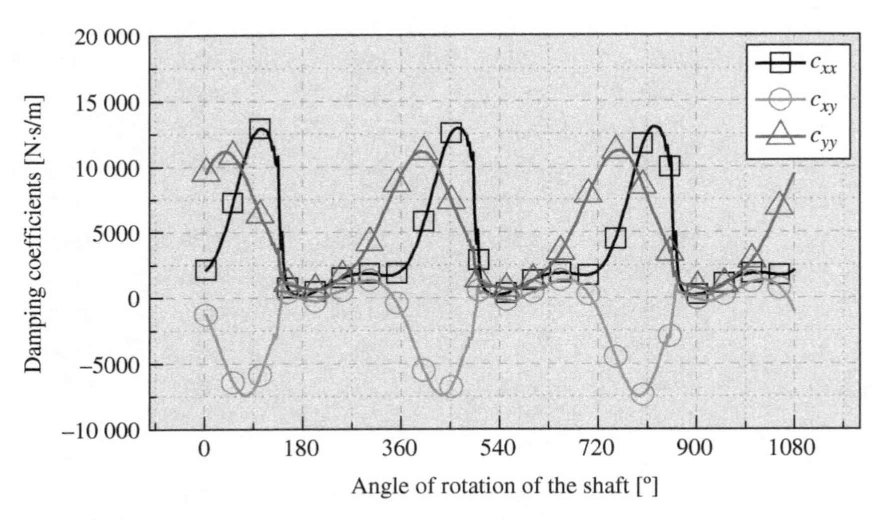

Figure 9.11 Damping coefficients calculated in the NLDW program for 3 rotor revolutions at a speed of 3250 rpm.

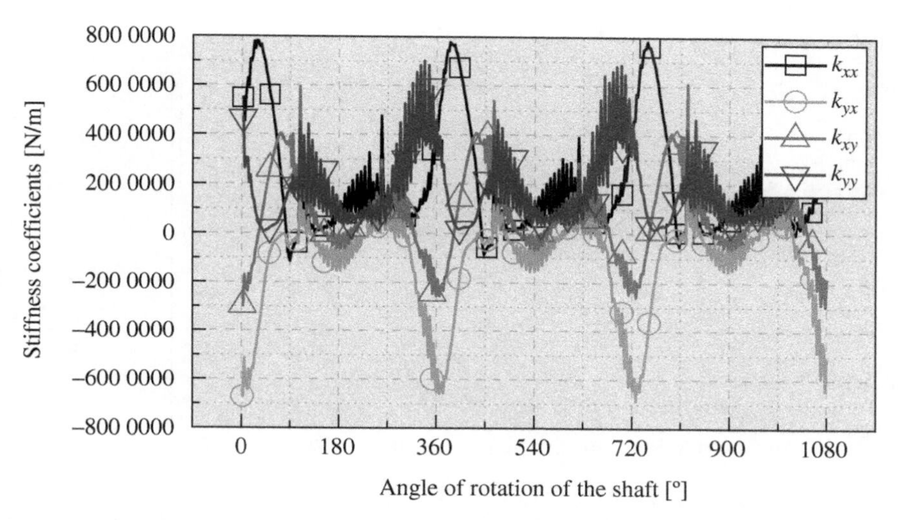

Figure 9.12 Stiffness coefficients calculated in the NLDW program for 3 rotor revolutions at a rotational speed of 3750 rpm.

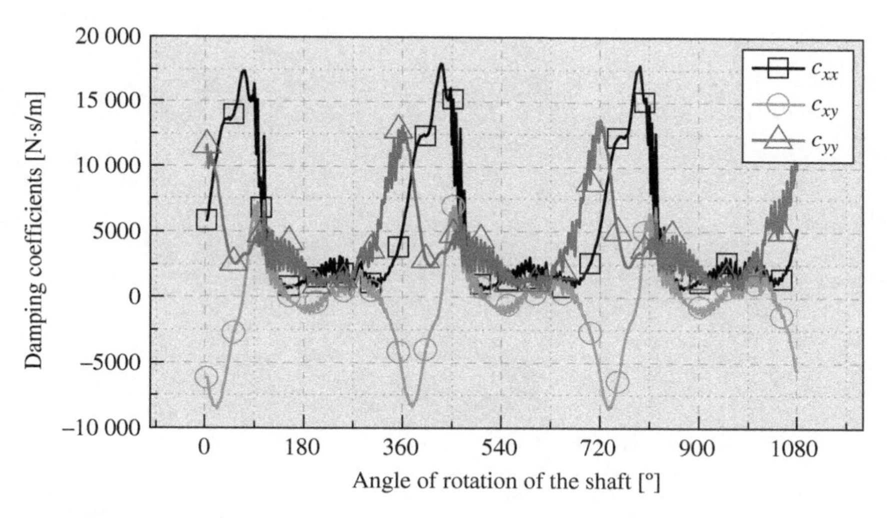

Figure 9.13 Damping coefficients calculated in the NLDW program for 3 rotor revolutions at a speed of 3750 rpm.

Calculated at 5500 rpm, the stiffness and damping coefficients look slightly different than in the case of the two previous described speeds. Changes in the stiffness and damping coefficients are not repeated according to the rotor speed. The stiffness coefficients in the *XX* direction change from −268 840 to 1 478 900 N/m, whereas stiffness coefficients in the *YY* direction change from −233 570 to 1 537 200 N/m. Values for the *XY* direction change from −2 520 300 to 43 052 N/m, whereas for the *YX* direction they change from 2 520 300 to 43 052 N/m. The damping coefficients in the *XX* direction change from −5047 N·s/m to about 588 N·s/m, whereas for the *YY* direction they change from −7046 N·s/m to approximately −560 N·s/m. The *XY* and *YX* diagonal coefficients have the same values and their values change from 1400 to 36 390 N·s/m.

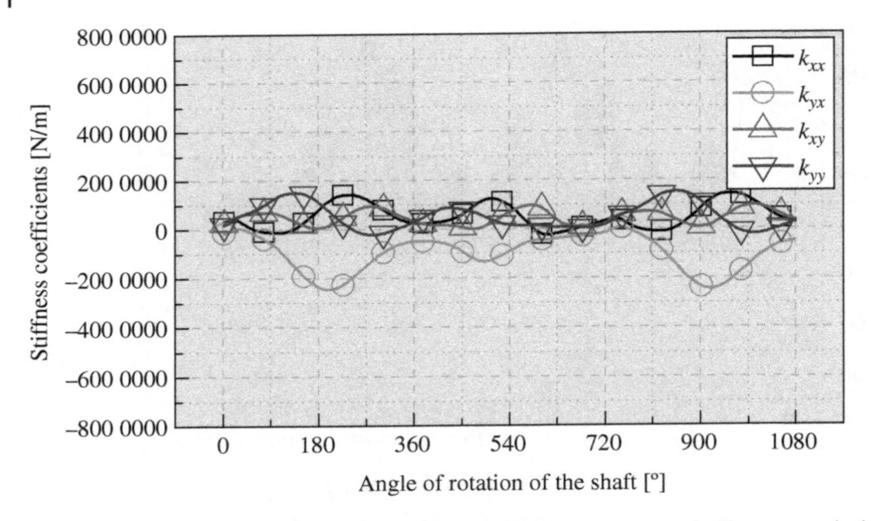

Figure 9.14 Stiffness coefficients calculated in the NLDW program for 3 rotor revolutions at a rotational speed of 5500 rpm.

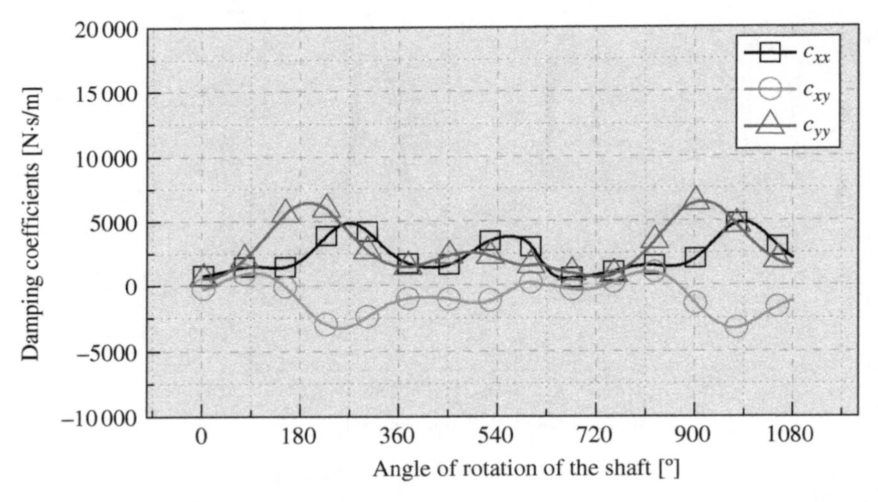

Figure 9.15 Damping coefficients calculated in the NLDW program for 3 rotor revolutions at a speed of 5500 rpm.

For all three speeds analyzed in the NLDW program, we observe very large changes in the stiffness sand damping coefficients as a function of speed. These changes are not repeated for each rotational speed within one rotor revolution, which is due to the fact that the rotor, in addition to the rotational movement, also performs a precession movement that causes such changes. A comparison of the results of calculations of dynamic coefficients of bearings calculated using linear and non-linear numerical and experimental methods is presented in Chapter 10.

9.4 Verification of Results Obtained

The stiffness and damping coefficients calculated on the basis of numerical calculation models are best verified on the basis of the rotor displacement obtained from them. The maximum displacements as a function of rotational speed and vibration trajectories can be directly and easily compared with the characteristics drawn up in Chapter 3.

Figure 9.16 shows the journal displacements in bearings no. 1 and 2 in the X and Y directions calculated using the LDW program on the basis of data from the KINWIR program. These results can be compared with the results of experimental studies presented in Figure 3.10. The resonance speed in experimental tests was about 4000–4500 rpm; the calculated speed in the KINWIR program is 3750 rpm. Amplitudes of resonant vibrations of the second bearing in the X and Y directions during experimental tests were about 140 and 120 μm, respectively, and for the first bearing their maximum value was 80 μm. Due to the influence of the fixed coupling, the resonance was not as visible as in the case of the results from bearing no. 2. The maximum journal displacements calculated in the KINWIR program in the second bearing were 130 and 170 μm. The vibration amplitude calculated numerically for the second bearing is approximately 90 and 130 μm. The results for the second bearing burdened with less influence of the coupling in experimental tests (in numerical studies the influence of the coupling was not taken into account) can be considered compatible.

The displacement of the journal in bearings no. 1 and 2 in X and Y directions calculated using the NLDW program (using a non-linear calculation algorithm) is shown in Figure 9.17. These results can be compared directly with the results of experimental studies presented in Figure 3.10. The curves measured during experimental tests and calculated numerically have a similar shape. In experimental tests, the highest vibration amplitude corresponding to resonant vibrations occurs at a rotational speed of 4500 rpm, whereas in

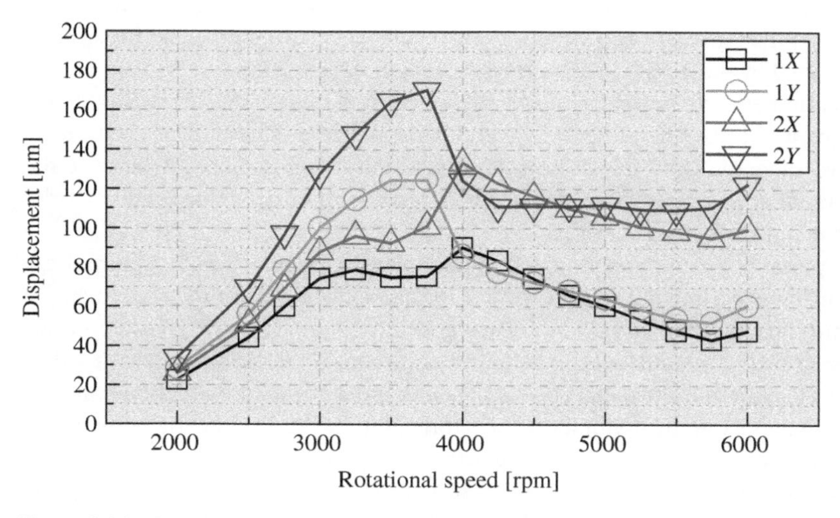

Figure 9.16 Bearing journal displacements calculated using the method with a linear calculation algorithm.

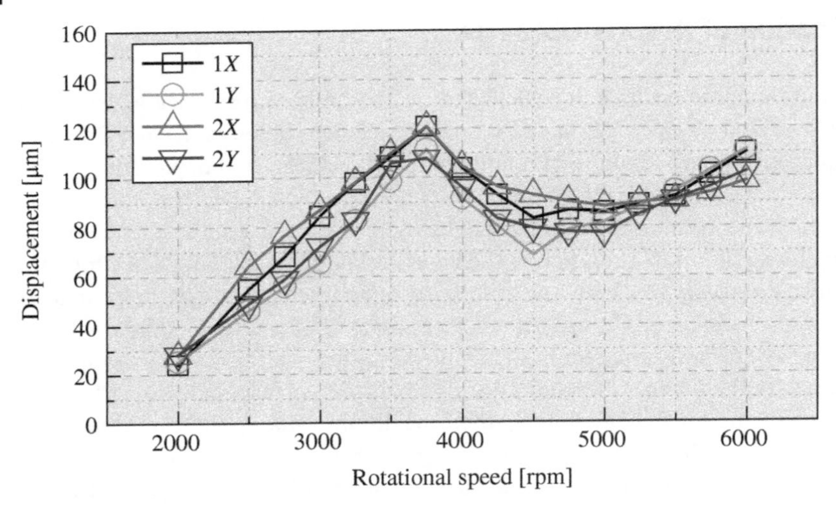

Figure 9.17 Bearing journal displacements calculated using the NLDW program.

numerical studies this value was found at 3750 rpm. Both by observing the results of experimental tests and numerical analysis, when the maximum vibration amplitudes are reached with the increase of rotational speed, this amplitude decreases to a speed of about 5000 rpm and then increases. The maximum amplitude of resonant vibrations of bearing no. 2 in experimental tests was about 120–140 µm. The resonant vibration amplitude of bearing no. 1 was lower due to the stiffening effect of the coupling and reached 80 µm. Calculated on the basis of numerical research, the amplitude of resonance vibrations for two bearings ranges from 110 to 120 µm. It is a value similar to that recorded during experimental tests for the second bearing, and at the same time this value is between the vibration amplitudes of the first and second bearings. Calculation results for a method with a non-linear calculation algorithm (in the NLDW program) reflect the results of experimental tests better than the results obtained by the method with a linear calculation algorithm.

Only the amplitudes of vibrations in the vicinity of bearing supports were measured during the experimental tests. Therefore, only a comparison of the vibrations of the bearing journals is presented. On the basis of numerical research it is possible to determine the rotor displacement at any point. The maximum displacement of the rotor center at resonance speed was approximately 1 mm.

Figure 9.18 presents a comparison of the vibration trajectories of bearing journal no. 2 measured during experimental tests (Figure 9.18a) and calculated numerically using a linear algorithm (Figure 9.18b) and a non-linear algorithm (Figure 9.18c). The results of this comparison described in more detail can be found elsewhere (Breńkacz and Żywica 2017). The vibration trajectories for the entire speed range measured during the experimental tests are shown in Appendix A. The vibration trajectories for the second bearing calculated numerically using a linear and a non-linear algorithm are shown in Appendix C. Figure 9.18 shows a comparison of three rotational speeds (3250, 4500, and 5750 rpm). Comparing the trajectories from the linear calculation model with the results of experimental tests, a similar shape and size can be observed, but it is "smoothed out". Comparing the results of non-linear analysis with experimental tests, a very high compatibility of the shape of the

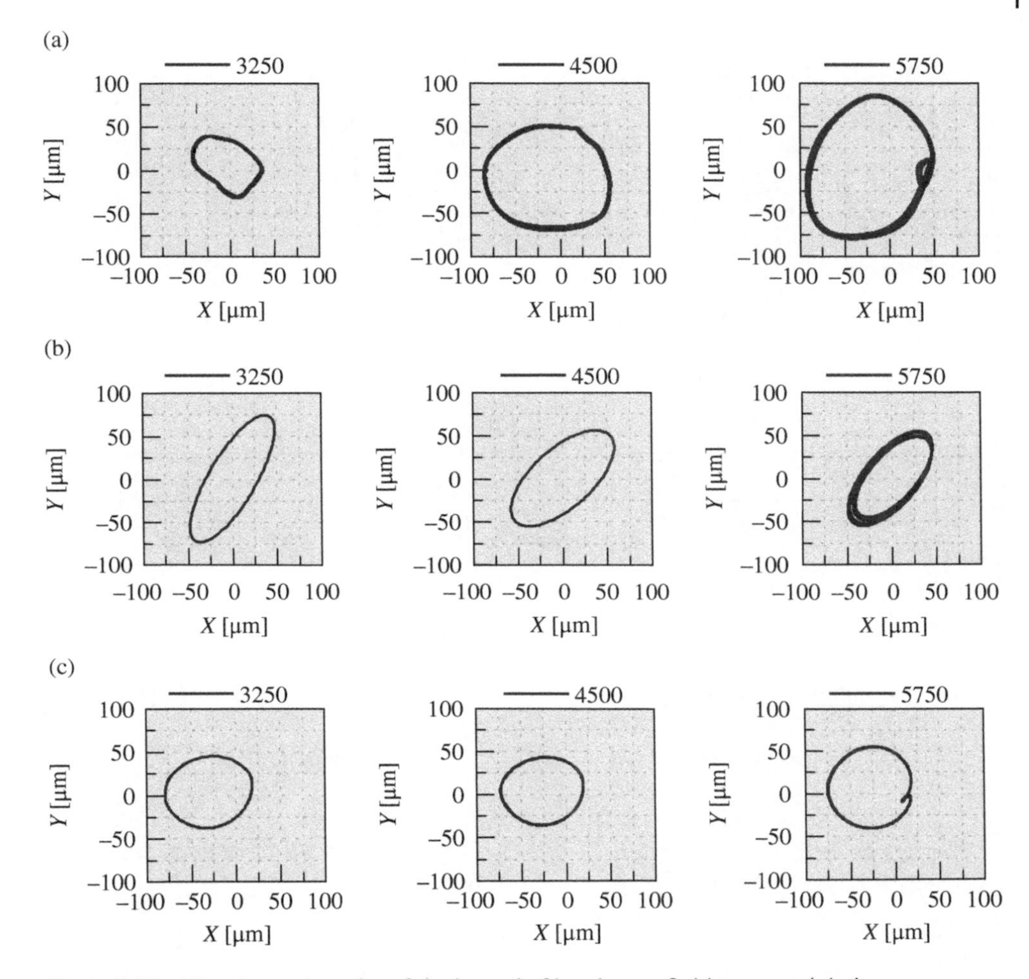

Figure 9.18 Vibration trajectories of the journal of bearing no. 2: (a) measured during experimental tests: (b) calculated using a linear algorithm: and (c) calculated using the NLDW program.

trajectory can be observed. Slight differences in numerical analysis results are visible in the size of the trajectory. These differences can be compensated for by adjusting the parameters of the numerical model, e.g. by increasing the unbalance of the rotor. Interestingly, the orbit calculated in the NLDW program for a rotational speed of 5750 rpm includes a small "loop" on the right side of the ellipse, and the result of the calculation is the same as in experimental tests. It is not possible to obtain such a vibration trajectory in linear calculation models.

Figure 9.19 presents the results of the fast Fourier transform (FFT) analysis for bearing journal vibrations no. 2 measured during experimental tests for a rotational speed of 5750 rpm in the X and Y directions (Figure 9.19a,b). The FFT analyses for the experimental signal for the entire speed range are presented in Appendix A.

Figure 9.19 presents the results of the FFT analysis for journal displacement signals generated on the basis of experimental tests as well as linear and non-linear numerical calculations. The Matlab program was used for the FFT analysis, similarly as in the case of the

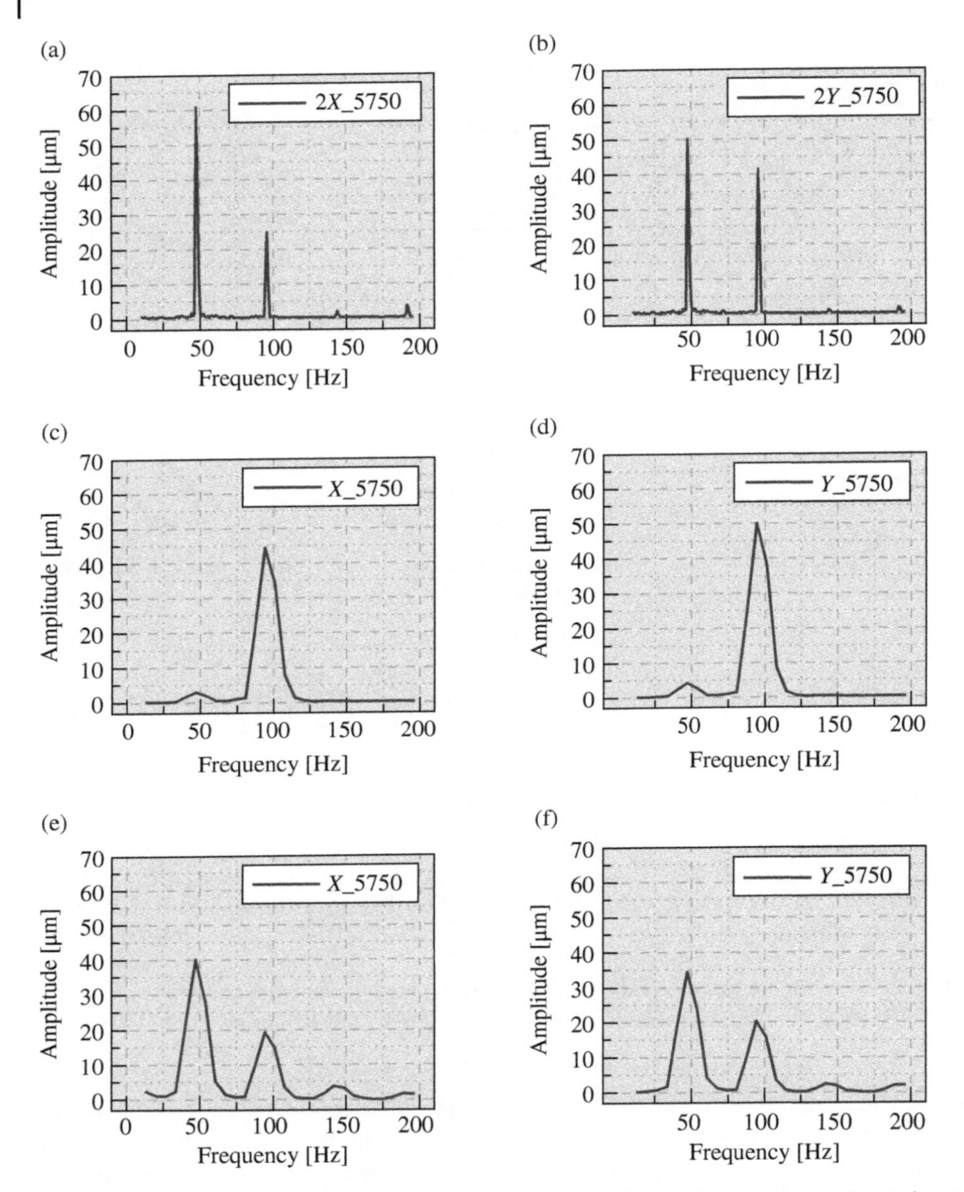

Figure 9.19 The FFT analysis of the signal for a speed of 5750 rpm based on experimental tests in the direction of (a) X and (b) Y; analogous results calculated numerically for the linear algorithm in the direction of (c) X and (d) Y; calculations by means of a non-linear algorithm in the NLDW program in the direction of (e) X and (f) Y.

signal obtained in experimental tests. When processing the results obtained numerically, the Hanning window was used, as described in Section 3.2. The different width of the curves in the FFT spectrum is caused by a different sampling rate of the signal.

Comparing the results of the linear analysis based on the stiffness and damping coefficients calculated in the LDW program and the experimental tests results (Figure 9.19a,b compared with Figure 9.19c,d), it is clear that approximately the same results are obtained. In the linear analysis there are no clear subsynchronous and super-synchronous components, they are, in turn, visible in the results of experimental tests. Calculated on the basis of numerical tests, the curve in the vibration spectrum associated with the synchronous speed of the rotor has a higher value than that obtained on the basis of experimental tests. In the X direction it is 45 μm instead of 25 μm, and in the Y direction it is 50 μm instead of 42 μm. The increase in this value may be due to the fact that there are no subsynchronous or supersynchronous speed components in numerical calculations.

Comparing the results of the non-linear numerical analysis carried out in the NLDW program with the experimentally measured values (Figure 9.19a,b compared with Figure 9.19e,f), it can be seen that the vibration spectrum has a very similar distribution in the X and Y directions. The height of the curves calculated numerically at approximately 54 and 108 Hz (synchronous speed) is 40 and 20 μm, respectively. The values obtained experimentally are 60 and 25 μm. Differences in curve heights result from differences in vibration trajectories presented earlier (in numerical calculations the vibration trajectories were smaller). The appearance of half-vibrations at about 54 Hz in the spectrum may indicate hydrodynamic instability of plain bearings.

9.5 Summary

The numerical calculations presented in this chapter were performed using a linear and non-linear model of numerical calculations. At the beginning of this chapter, the procedure for calculation of the stiffness and damping coefficients was presented. The selected method of calculation assumes the use of the perturbation method. This method is based on the development of all components of the Reynolds equation into Taylor's series at the point of static equilibrium. The linear and non-linear calculation algorithms were presented. In the linear calculation model it was necessary to assume small changes in vibration amplitude and calculations were performed for the static equilibrium point. In this case, one set of stiffness and damping coefficients for one speed was obtained. For each speed there are 8 coefficients: 4 stiffness coefficients (2 main and 2 cross-coupled) and 4 damping coefficients (2 main and 2 cross-coupled). The cross-coupled damping coefficients have the same value. The dynamic coefficients of bearings changed with the rotational speed changes.

In the non-linear case, there are large journal displacements within the lubricating gap. To calculate such a case in the perturbation method, the Reynolds equation was solved together with the dynamic component. For small time intervals it is possible to find sufficiently small movements in which the properties of the oil film are constant. In the non-linear calculation it is also necessary to take into account the pressure in the bearing not at the static equilibrium point, but for the momentary position of the journal in the lubrication gap. In order to accurately calculate the bearing stiffness and damping coefficients (taking into account the influence of temperature), in addition to solving the Reynolds equation, it is also necessary to solve the energy equation, the conductivity equation, the equation with the description of the lubrication gap shape, and the disturbing equations. All of the above equations are linked to each other and their solution is time-consuming.

A rotor and bearing model was used for the calculations, reflecting the experimental tests carried out earlier, so that it was possible to directly compare the results obtained on the basis of numerical analyses and experimental tests. The chapter presents the results of numerically calculated stiffness and damping coefficients

The non-linear calculation shows the extent to which the values of the dynamic coefficients calculated for one speed can change. During rotor operation at resonance speeds, the change in dynamic coefficients of bearings was greater than during rotor operation at other rotational speeds. Observing the data from the numerical non-linear model it can be seen that for some rotational speeds, changes in stiffness and damping coefficients occur cyclically with the rotor revolutions, however for some rotational speeds these changes are more irregular. This is due to the fact that the rotor not only makes a rotary motion, but also a precession motion, when it is lifted by the lubricating film layer. Making a vast generalization, it can be stated that the greater the change in stiffness coefficients is related to the complex operation of the bearing. If, in addition to the harmonic components, high subharmonic and superharmonic components are observed in the vibration spectrum, this means an unstable bearing operation. This was observed for a wide range of tested rotational speeds.

The obtained calculation results were verified. On the basis of the obtained stiffness and damping coefficients in the linear and non-linear numerical model, journal displacements generated for a wide range of rotational speeds were calculated. On this basis, vibration trajectories and the FFT diagrams of the obtained signals were also created. In the linear model a correct convergence of results was achieved, the vibration trajectories have a similar amplitude, but they are "smoothed out." In the non-linear model, a very good trajectory convergence was achieved. In non-linear numerical calculations, in addition to the main trajectory, smaller "loops" were also obtained. Numerically calculated journal displacements in bearings are approximate to those obtained experimentally. The results of non-linear analysis better reflect the tested real system.

Calculation of the vibration trajectory based on the linear (fixed) values of the stiffness and damping coefficients allows the shape of the main ellipse of vibrations to be reproduced, but it is not possible to accurately reflect the real trajectories (taking into account that we analyzed system working with strong nonlinearity). Comparing the FFT diagrams of vibrations generated by means of a numerical analysis with those recorded during experimental tests also shows the correctness of the results of numerical calculations. In this case, the results of linear tests should also be considered correct, and the results of non-linear tests reflect the results of experimental tests with high accuracy. A very accurate reflection of the experimental results is the best proof of the correctness of the calculated stiffness and damping coefficients.

10

Comparison of Bearing Dynamic Coefficients Calculated with Different Methods

An algorithm of calculation of dynamic coefficients of hydrodynamic bearings on the basis of experimental tests was developed. Verification of the method and analysis of its sensitivity was carried out, during which the influence of various factors was checked, such as a different way of inducing vibrations by means of a impact hammer, influence of uneven distribution of force between bearings and asymmetrical rotor on the values of calculated dynamic coefficients of bearings. In this way it was possible to determine which parameters are important. In the course of the work it turned out that most of the factors that could potentially affect the results in a negative way do not have a great influence on them, and a large number of them are easy to correct. The method of making the necessary adjustments and their impact on the calculated values was described.

In addition to the verification of the calculation method itself, the results obtained from experimental tests were also verified using a numerical model developed in the Abaqus program. A mass equal to half the mass of the rotor has been assigned to a point with one degree of freedom. This point was connected to the second fixed point by means of an elastic-damping element. The values of stiffness and damping in the numerical model prepared in such a way were set as values calculated in experimental tests. A force of the same value as that measured during experimental tests was applied to the mass point. It turned out that the mass point displacement obtained on this basis corresponds to the displacement recorded during the experimental tests. On this basis, it can be concluded that the experimental method can be successfully used to determine the dynamic coefficients of bearings.

On the basis of the results obtained using the experimental method of determining dynamic coefficients of bearings (applied in a wide range of rotational speeds), it was found that changes in the calculated coefficients as a function of rotational speed run smoothly with the increase in rotational speed. It is important to know that the analyzed system operates with strong nonlinearity. Obtained curves were as expected and standard deviations were quite high but still at an acceptable level. Of course, as the standard deviation increases, the variability of the determined dynamic coefficients of bearings for a given speed is greater, and in the ranges where the values were calculated with a smaller standard

Bearing Dynamic Coefficients in Rotordynamics: Computation Methods and Practical Applications, First Edition. Łukasz Breńkacz.

Companion website: www.wiley.com/go/brenkacz/bearingdynamiccoefficients

deviation, more unambiguous results were obtained. At lower speeds (below resonance speed), the results were characterized by a smaller standard deviation.

Numerical calculations were carried out using linear and non-linear calculation algorithms. The results obtained on the basis of numerical calculations were verified by comparing the maximum displacements of bearing journals, journal trajectories and the fast Fourier transform (FFT) diagram with the results of experimental tests. The stiffness and damping coefficients obtained from linear numerical research can be successfully used for the initial estimation of rotating machinery dynamics. The results calculated with a non-linear algorithm of stiffness and damping coefficients allow for a very accurate reflection of the rotor dynamics. The calculated vibration trajectories take into account the extremely complex shape of the trajectory, which was observed during the experimental tests.

After all these positive verifications of the results obtained on the basis of experimental and numerical tests, it is puzzling why the obtained dynamic coefficients of bearings are very different from each other. The results obtained from linear and non-linear numerical analysis and experimental method for a speed of 3000 rpm are presented in Table 10.1.

Figure 10.1 presents a proposed explanation for obtaining such different results. This is also the main difference between linear and non-linear numerical calculations and experimental tests. This is the way the system works in the lubrication gap, for which the stiffness and damping coefficients were determined. Numerical analyses based on linear algorithms for calculating stiffness and damping coefficients are carried out for the static equilibrium point denoted by O_c in the figure. For this point, the reactions of the lubricating film are determined, and on their basis the stiffness and damping coefficients are determined. In practice, however, larger vibration trajectories in hydrodynamic bearings can be often observed. For such cases it is necessary to use non-linear algorithms in numerical calculations.

Examples of increasing vibration amplitude are shown in Figure 10.1a – orbits denoted by 1, 2, and 3. If the experimental method is used to determine the trajectory of the rotor working at the point of static equilibrium, an example of the trajectory shown by a dotted line in Figure 10.1b after the excitation force can be observed. If experimental tests are carried out during the operation of the rotor, e.g. on trajectory 2, the result obtained after the

Table 10.1 Comparison of the calculated stiffness and damping coefficients for the three methods used for a speed of 3000 rpm.

Coefficients	Numerical method with a linear algorithm	Numerical method with a non-linear algorithm	Experimental method
k_{xx} (N/m)	369 000	From −203 540 to 1 782 500	5989 ± 3610
k_{xy} (N/m)	38 700	From −265 700 to 1 050 100	−4918 ± 9423
k_{yx} (N/m)	−534 000	From −2 087 900 to 98 416	13 801 ± 7539
k_{yy} (N/m)	354 000	From −228 470 to 1 326 500	3308 ± 7857
c_{xx} (N·s/m)	2310	From 212 to 8781	56 ± 22
c_{xy} (N·s/m)	−1420	From −4885 to 629	12 ± 14
c_{yz} (N·s/m)	−1430	From −4885 to 629	28 ± 11
c_{yy} (N·s/m)	2060	From 462 to 7629	36 ± 22

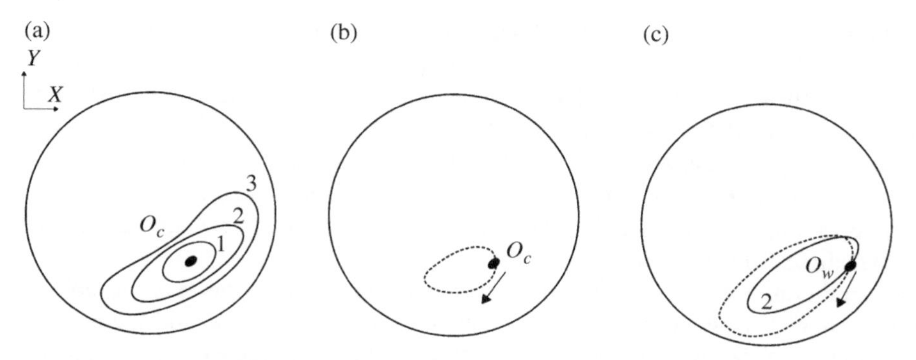

Figure 10.1 (a) Change in journal vibration trajectory due to e.g. increased rotor unbalance or changes in bearing parameters (trajectory 1, 2, or 3). (b) Trajectory after inducing a bearing operating at a static equilibrium point (O_c). A model showing the assumptions of linear operation of the system. (c) Trajectory after excitation during operation with a larger ellipse of vibrations. The dotted line shows a sample trajectory after excitation in point O_W. After a short time, the rotor journal returns to the previous constant ellipse on which it was moving earlier (shown by a continuous line).

excitation is similar to the one shown in Figure 10.1c. The dotted line shows the displacement of the rotor after excitation in point O_W. After using a small amount of excitation force, the rotor moves along the trajectory shown by the dotted line, while after a short time the rotor returns to stable operation and moves along the trajectory 2 again. At each point of the trajectory created in this way, the properties of the lubricating film change, and using an experimental method of identifying dynamic coefficients of bearings, the parameters that are the result of shaft journal displacement are determined.

Calculations by all three methods generate different results due to the fact that the calculations are carried out for different rotor journal positions. In the calculations carried out using the linear calculation method obtained from experimental tests, it must be assumed that the rotor operates at a static equilibrium point, and the excitation starts at the static equilibrium point, and after a short time and small displacements of the bearing journal it also returns to the static equilibrium point. In practice, however, if the rotor journal moves along a relatively large trajectory and vibrations are induced, then moving along the new trajectory created in this way and changing its position in subsequent moments of time, it will occupy successively places in the lubrication gap of a bearing with different properties of a lubricating film. The problem in this case does not seem to be the magnitude of the vibration trajectory itself, but the high variability of the lubricating film properties with small changes in the displacement of the journal. Theoretically, assuming that in the whole lubrication gap the values of stiffness and damping coefficients of the lubricating film are similar, the determination of dynamic parameters of bearings on the basis of experimental tests would make it possible to determine the stiffness and damping coefficients with the same values as those calculated on the basis of numerical analyses.

It is the large changes in the lubricating film properties for small changes of the journal position that cause differences between the experimentally and numerically determined stiffness and damping coefficients and are the main limitation in the use of experimental methods for determining the dynamic properties of rotating machinery.

The differences between the results of linear and non-linear numerical analysis can be explained by applying a similar argument to the one presented earlier in the case of experimental tests. In the case of numerical analysis by means of a linear algorithm, it shall be assumed that the calculations are performed for a static equilibrium point. In the case of non-linear analysis, the trajectory is divided into sufficiently small fragments, assuming constant values of stiffness and damping coefficients. In an ideal linear case, when this bearing is operating at a static equilibrium point, the results of linear and non-linear numerical analyses will be the same.

During operation, it turned out that the operation of the bearings was very unstable in the tested system. The generated vibration trajectories were quite large, and the bearing characteristics were clearly non-linear. In the results of the FFT analysis for most of the rotational speeds the components associated with very high (often twice as high as synchronous vibrations) subsynchronous and supersynchronous vibrations were visible. This case proved to be ideal for verifying differences in the results of stiffness and damping coefficients calculated using different methods, as differences in the results obtained are very evident.

In operating practice, in a properly working machine, bearings should be designed in such a way as to ensure stable rotor operation. The easiest change that will improve dynamic properties is to change the geometry of the bearing lubrication gap. Already after preliminary analyses, a vast improvement in dynamic parameters of bearing operation will be obtained by reducing the size of the lubrication gap. However, it should be kept in mind that when reducing the size of the lubrication gap, attention must be paid to problems related to manufacturing precision and increased temperature during bearing operation. If the dynamic coefficients of bearings are calculated using three methods for the case where the rotor operates close to the static equilibrium point, all methods will allow similar results to be obtained.

11

Summary and Conclusions

This book is devoted to the study of dynamic properties of rotating machinery by means of numerical and experimental calculation methods. Dynamic properties of rotating machinery are understood to be the stiffness, damping and mass coefficients of the rotor–bearing system. The main aim of the work was to develop an experimental method of determining the dynamic coefficients of radial hydrodynamic bearings and to calculate the values of dynamic coefficients of bearings on the basis of experimental tests as well as linear and non-linear numerical methods. Experimental tests were carried out on a laboratory test rig with a rotor supported on two hydrodynamic radial bearings.

The tests carried out in this monograph are shown in Figure 11.1. Experimental tests were complemented by numerical analyses at several stages of research. The main stream of research was connected with experimental and numerical determination of dynamic coefficients of bearings. Chapter 3 and Appendix A present the basic characteristics of the laboratory test rig, e.g. run-up and run-out characteristics of the rotor, vibration trajectories, the fast Fourier transform (FFT) analysis of signals, and modal analysis of the test rig construction. Chapters 5–8 describe the process of experimental determination of dynamic coefficients of bearings. A method with a linear calculation algorithm with impulse excitation was used. Chapter 5 presents the calculation algorithm. Chapter 6 describes in more detail one of the signal processing stages, i.e. operations aimed at eliminating the impact of rotor unbalance on the calculation results. Chapter 7 describes the sensitivity analysis of the method. The influence of six parameters on the calculated results of dynamic coefficients of bearings and ways of improving the results were checked. The course of experimental tests is presented in Chapter 8, which shows the signals of excitation forces and system responses, as well as the results obtained. The Matlab program was used for signal processing and calculation of dynamic coefficients of bearings.

Samcef Rotors was used to verify the algorithm used for experimental determination of dynamic coefficients of bearings and analysis of its sensitivity (Chapters 5–7). Verification of the results obtained from experimental tests (described in Section 8.7) was carried out in the Abaqus program using a numerical model with one degree of freedom. A material point with a mass equal to half that of the rotor is fixed to a second fixed point by means of an elastic-damping element. The values of stiffness and damping coefficients calculated

Bearing Dynamic Coefficients in Rotordynamics: Computation Methods and Practical Applications, First Edition. Łukasz Breńkacz.

Companion website: www.wiley.com/go/brenkacz/bearingdynamiccoefficients

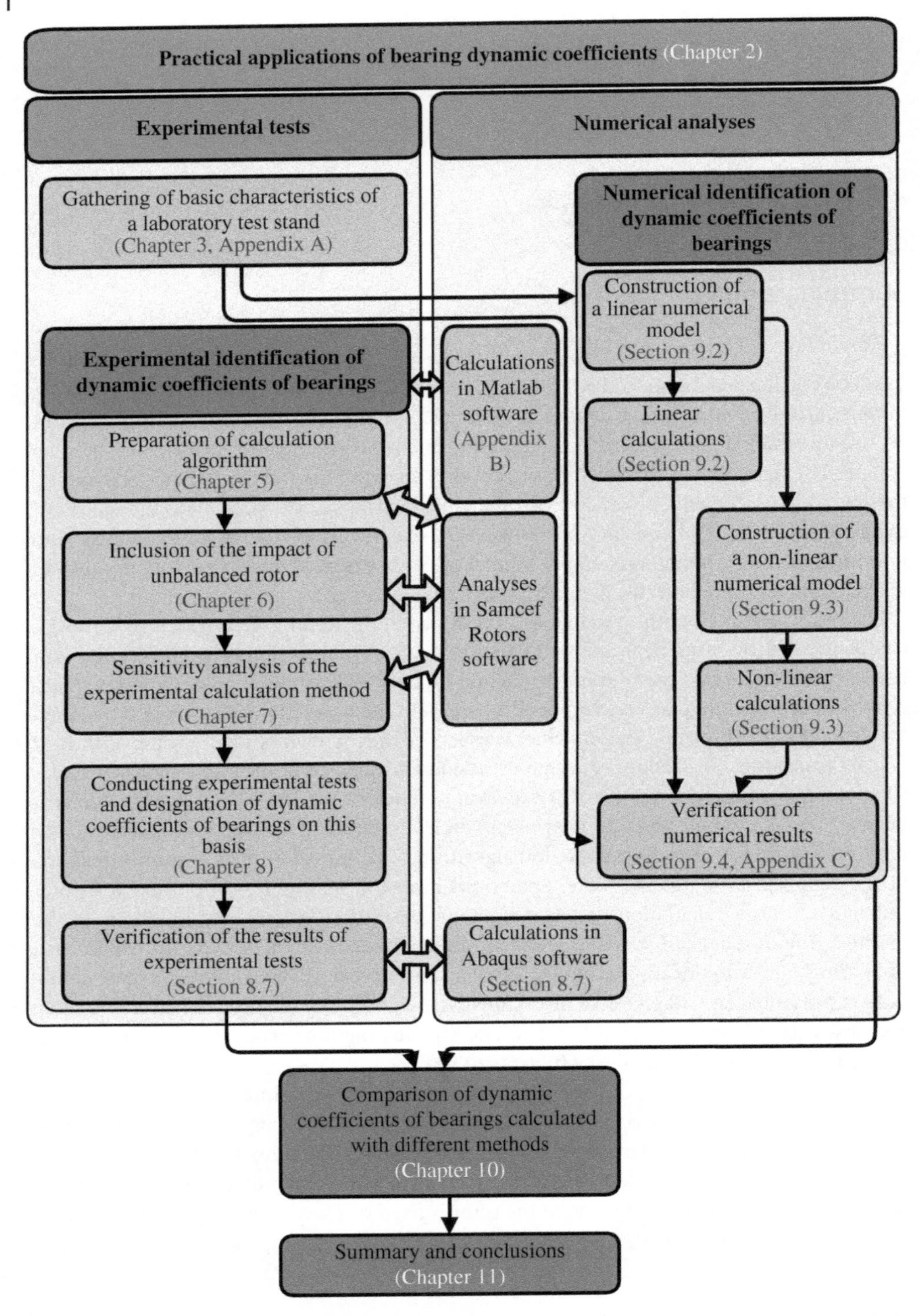

Figure 11.1 A diagram of the work carried out. The division into experimental tests and numerical analyses and the relationships between the various chapters and sections of the monograph are indicated.

based on experimental tests were assigned to the elastic-damping element. The material point was induced to vibrate by an impulse force. The force from experimental tests was used for the analysis, on the basis of which the stiffness and damping parameters were determined experimentally. After comparing the displacements of the material point calculated in the Abaqus program and those measured during experimental tests, it turned out that the results obtained were convergent. This proves the correctness of the experimentally determined stiffness and damping coefficients.

An interesting result is the observation of changes in the values of mass coefficients calculated on the basis of experimental tests together with the change of rotational speed. The mass coefficients can be interpreted as the mass of the part of the shaft involved in vibrations. It can be seen that at low speeds (2250 rpm) the mass coefficient in the first main direction m_{xx} has a relatively low value (−0.4 kg), the second main mass coefficient m_{yy} has approximately the same positive value. When the changes in this factor are approximated with linear or low-order polynomial functions, it can be seen that as the rotational speed increases, the first factor increases and the second factor decreases. At higher speeds (6000 rpm) these values are almost the same as at lower speeds, but with opposite signs. Resonance occurs at the point where lines that approximate mass coefficients pass through the zero value. However, for the resonance speed alone, the values of these coefficients do not equal zero. The mass of the shaft, estimated on the basis of calculated mass coefficients for the second bearing and a speed of 2250 rpm, is 1.18 ± 0.56 kg. Comparing this value with the expected value, i.e. half of the weight of the shaft (i.e. 2.35 kg), the result is about twice as low as the expected value.

Numerical calculations were carried out using the KINWIR, LDW (linear calculations) and NLDW (non-linear calculations) programs. A description of the numerical model and the results of the calculations are presented in Chapter 9. On the basis of the calculated stiffness and damping coefficients, the displacements of the bearing journals and their vibration trajectories were determined. The results of numerical calculations were compared with the results of experimental tests. In this way, the analysis of the correctness of the results of calculations of dynamic coefficients of bearings was carried out using numerical methods. Correct results of linear numerical calculations were obtained. In the case of numerical calculations using a non-linear calculation algorithm, very accurate trajectories were obtained, taking into account the complex shape recorded during experimental tests. The convergence of the results indicates the correctness of the non-linear form (functions variable in time) of the stiffness and damping coefficients.

The results of calculations of stiffness and damping coefficients using different methods are presented in Chapter 10. The results of calculations obtained using an experimental method based on linear identification algorithm and numerical methods were compared with linear and non-linear calculation algorithms. It should be stressed that the calculations were made on a system characterized by non-linear operating parameters (there were subsynchronous and supersynchronous components in the rotor's vibration spectrum for almost the entire range of rotational speeds). Linear and non-linear numerical analyses of a rotating system operating at a static equilibrium point (an idealized case that in reality is non-existent) show that in the case of two numerical methods similar results are obtained. From a scientific point of view, it is interesting to carry out linear and non-linear calculations on a real system operating in a "visibly non-linear" manner. In practice, all rotating systems are characterized by such properties (they are more or less non-linear). It turned out that linear numerical calculation methods, commonly used for the initial calculation of hydrodynamic plain bearings, determine only approximate values of the bearing stiffness and damping parameters.

The values of stiffness and damping coefficients of hydrodynamic bearings calculated on the basis of experimental studies were verified at several stages. The results of numerical analyses carried out using linear and non-linear numerical algorithms were also verified. After comparing the results, it was found that the results obtained in all three cases are different. An explanation for this fact is proposed in Chapter 10. The reason is that the coefficients are determined in all three methods for different positions of the rotor journal. In linear numerical research, it is the position at the point of static equilibrium. For non-linear numerical testing, calculations are made for the subsequent rotor journal positions. In experimental tests, dynamic coefficients of bearings are determined on the basis of trajectories other than those obtained during stable operation.

It is also worth noting that the presented work includes original elements. The most important ones include:

1) Descriptions are given of practical applications of bearing dynamic coefficients.
2) Development of a method for experimental determination of all dynamic parameters for two hydrodynamic radial bearings in one operation. Compared with the methods described above it was extended by the possibility of determining mass coefficients, which enables quick verification of the obtained results.
3) Verification of the tested algorithm was carried out on the basis of a numerical model of a rotating machine created in the Samcef Rotors software. It turned out that there is a possibility of experimental determination of mass, stiffness and damping coefficients with very high theoretical accuracy – usually with error not exceeding 1%.
4) A sensitivity analysis of the experimental method of determining the stiffness, damping and mass coefficients of hydrodynamic bearings was carried out. The influence on the results of parameters such as excitation force location, excitation angle, unbalance, displacement measurement location, rotor asymmetry and for the calculation of a rotor with different Young's modulus has not been studied before.
5) Dynamic coefficients of bearings were calculated on the basis of experimental tests over a wide range of rotational speeds, taking into account resonance and ultra-resonance speeds. Previously published analyses were usually limited to a single speed or, in rare cases, to a narrow speed range, where the stiffness and damping coefficients did not differ significantly from each other.
6) A comparison of dynamic coefficients of bearings calculated on the basis of three different calculation methods – an experimental method based on a linear calculation algorithm as well as numerical methods based on a linear and a non-linear calculation algorithm – was carried out. Each method was verified. The obtained calculation results were decisively different. The reason for these differences was discussed.

Concluding the summary of this work, the subject matter tackled in this monograph should be developed further, namely the range of applicability of linear methods for determining dynamic properties of bearings should be determined. Comparing the results of calculations for the non-linear system, obtained using two linear methods (experimental and numerical) and a non-linear numerical model, it can be stated that the use of all three calculation methods described is justified in different cases. If the numerical model of bearings, rotor and supporting structure is well defined, the numerical non-linear calculation model makes it possible to obtain accurate values of dynamic coefficients of bearings. In practice, each rotating system exhibits non-linear properties, therefore this method should be applied in a wide range of cases. In cases where quick calculations are required, or if only linear calculation models are available, numerical methods based on linear algorithms can be successfully used for carrying out initial analyses. A numerical method with a linear calculation algorithm can be used for non-linear systems and it is possible to obtain results similar to the real results on this basis. When dealing with a very complex structure of dynamic systems or when not all parameters of bearings and supporting structure are known, it is reasonable to use experimental methods for determining stiffness and damping coefficients of bearings. Experimental calculation methods based on linear algorithms should only be used in linear systems. Their use in non-linear systems (large displacements with large changes in lubricating film properties) may provide results burdened with large errors.

In the case under consideration (a strongly non-linear system) the results obtained by means of the experimental method of determining dynamic coefficients of bearings can only be used as a reference. On the basis of negative damping, it is possible to evaluate, for example, the place of resonance occurrence, and the courses of changes in stiffness, damping and mass coefficients characterize well the changes in the dynamics of the system due to impulse excitation. However, the dynamic coefficients of bearings, obtained from this analysis, cannot be directly used for further calculations of the dynamics of the structure.

One of the greatest advantages of the impulse method of experimental determination of stiffness and damping coefficients described in this work is its versatility. The fact that this method can be used directly for different types of bearings is important. Tests on hydrodynamic bearings made it possible to compare the results with very complex linear and non-linear numerical calculation models. The experimental method can be used in the calculation of dynamic coefficients of bearings in cases where not all parameters of the bearings are known. There are new types of bearings for which the parameters are extremely difficult to determine using numerical calculation, e.g. magnetic bearings (Zhou et al. 2016), segmented bearings, hybrid bearings, or foil bearings.

The author hopes that the experimental method of determining dynamic coefficients of bearings described in this work and analysis thereof will contribute to a wider application in practice, at least to a small extent. Knowing the possibilities of the experimental method (theoretically, calculations can be carried out with very high accuracy) and the associated limitations (use in non-linear systems is associated with a calculation error), it can be used as a valuable research tool.

Appendix A

This appendix presents the results of experimental tests to complement and extend those contained within the monograph.

A.1 Fast Fourier Transform Diagrams of Journal Displacement in Relation to Bearing Housings

The method of creating diagrams is described in Section 3.1. Figures A.1–A.4 describe the results showing 64 diagrams for two bearing supports in the *X* and *Y* directions made for 16 rotational speeds: from 2250 to 6000 rpm (in steps of 2250 rpm).

A.2 Journal Vibration Trajectories in Relation to Bearing Housings

The method of creating diagrams is described in Section 3.1. Figures A.5 and A.6 show 32 diagrams. Each represents 12 revolutions of the rotor.

A.3 Acceleration of Vibrations of Bearing Supports during Start-up

The interpretation of the cascade diagram is presented in Section 3.2. Figures A.7–A.10 present the signal recorded during the run-up of the laboratory test rig, from 1000 to 8400 rpm. The signal was recorded by means of accelerometers placed in two bearing supports in directions perpendicular to the shaft axis.

Bearing Dynamic Coefficients in Rotordynamics: Computation Methods and Practical Applications, First Edition. Łukasz Breńkacz.

Companion website: www.wiley.com/go/brenkacz/bearingdynamiccoefficients

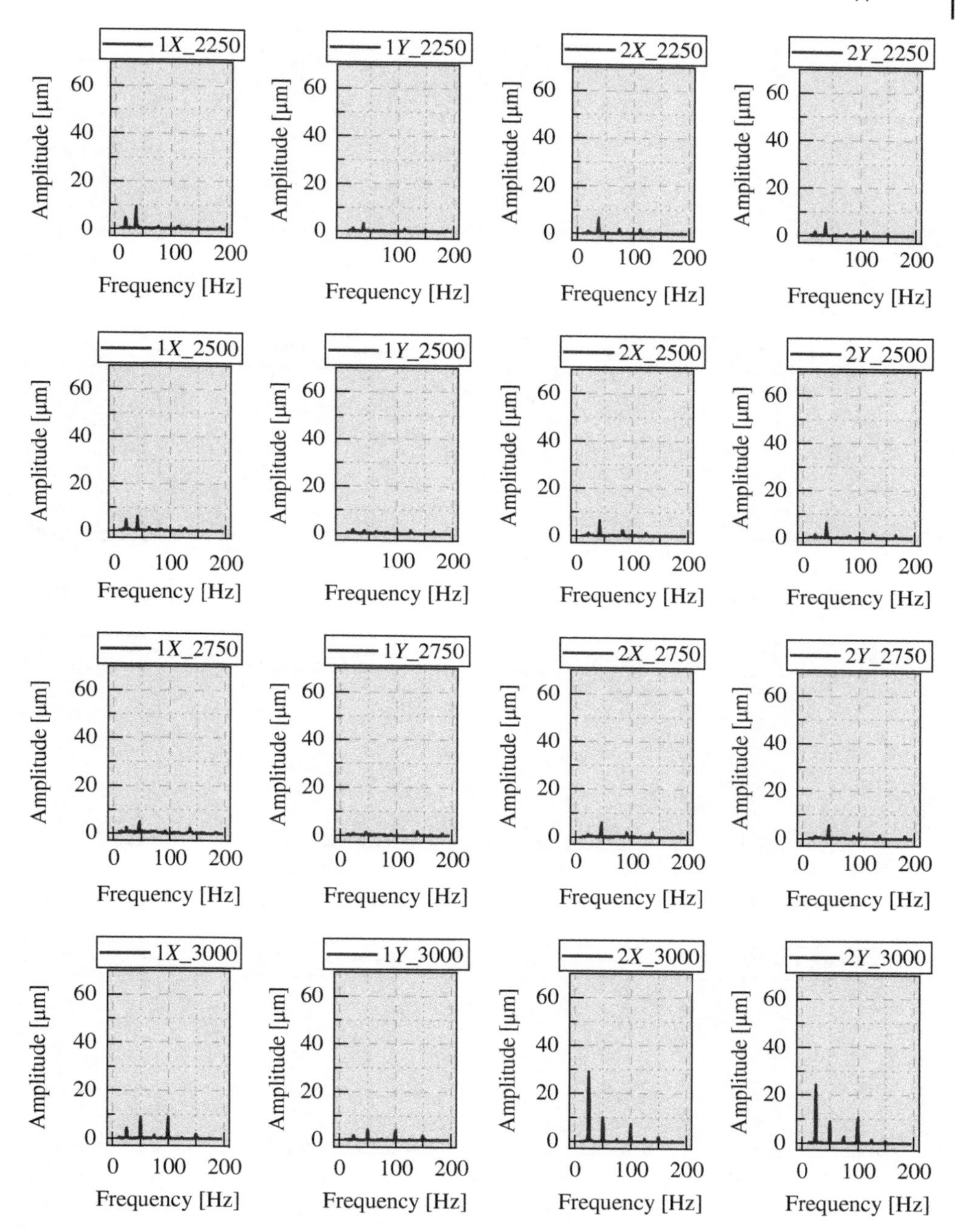

Figure A.1 Fast Fourier transform (FFT) diagrams of the first and second bearings in the X and Y directions for rotational speeds of 2250–3000 rpm.

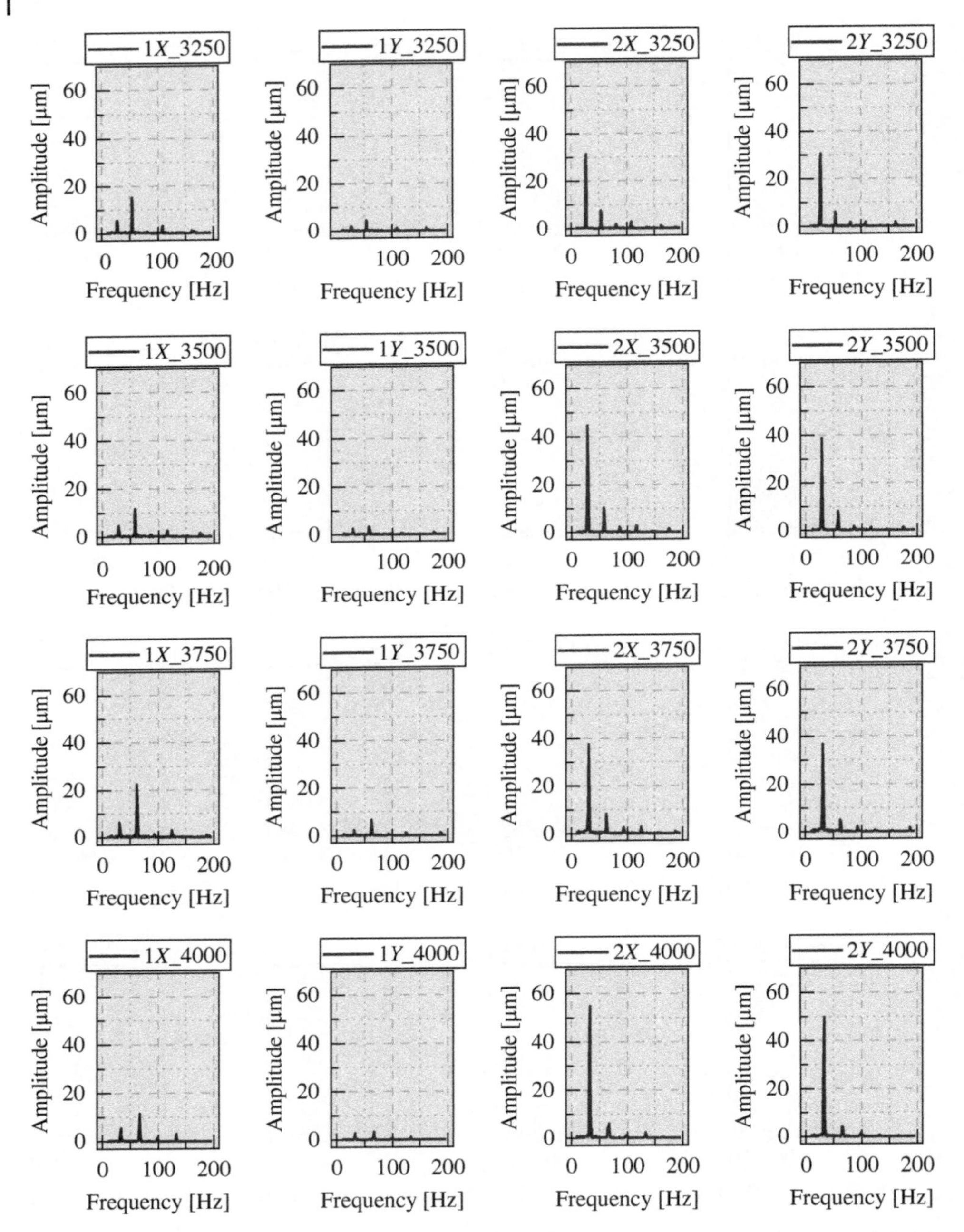

Figure A.2 FFT diagrams of the first and second bearings in the X and Y directions for rotational speeds of 3250–4000 rpm.

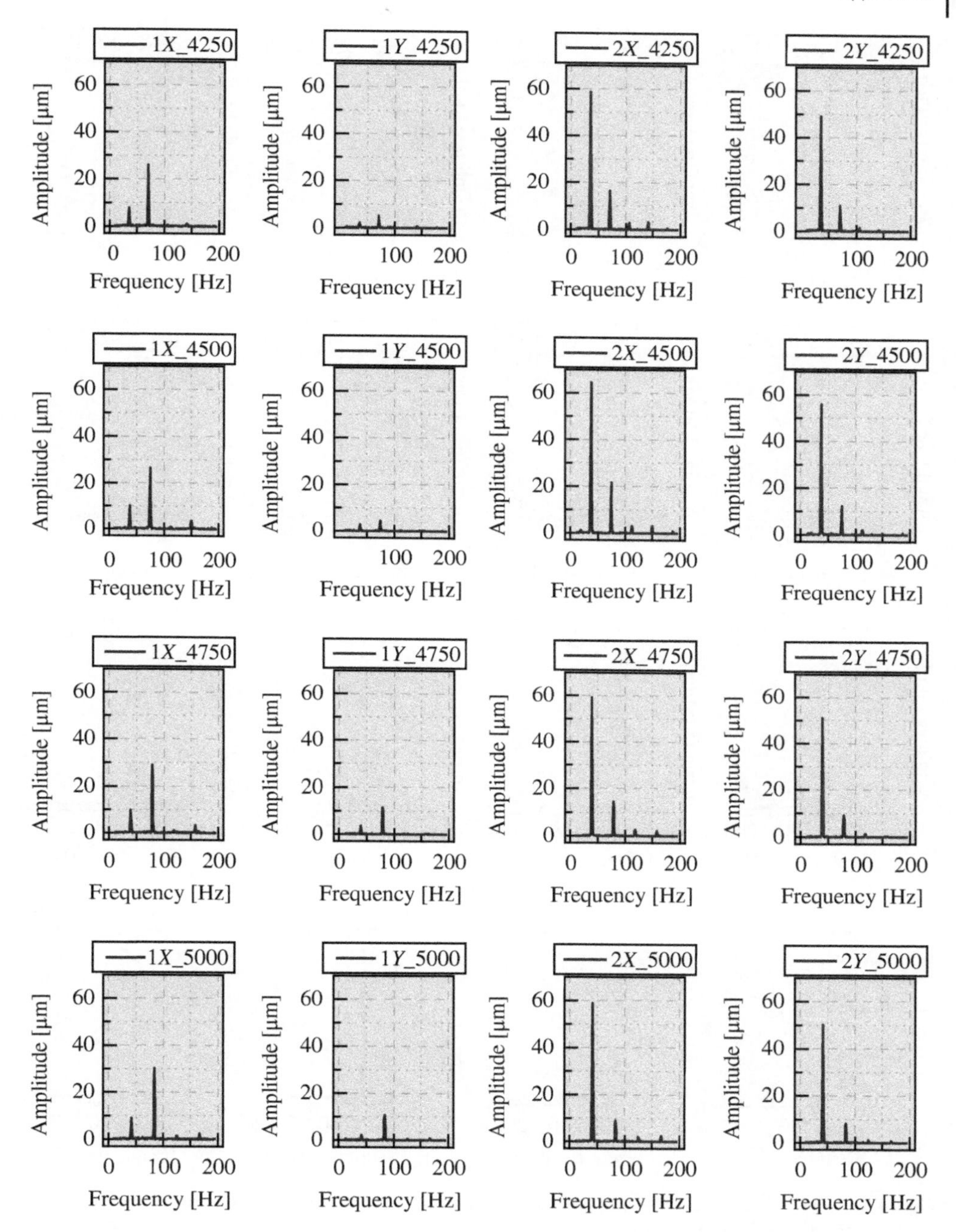

Figure A.3 FFT diagrams of the first and second bearings in the X and Y directions for rotational speeds of 4250–5000 rpm.

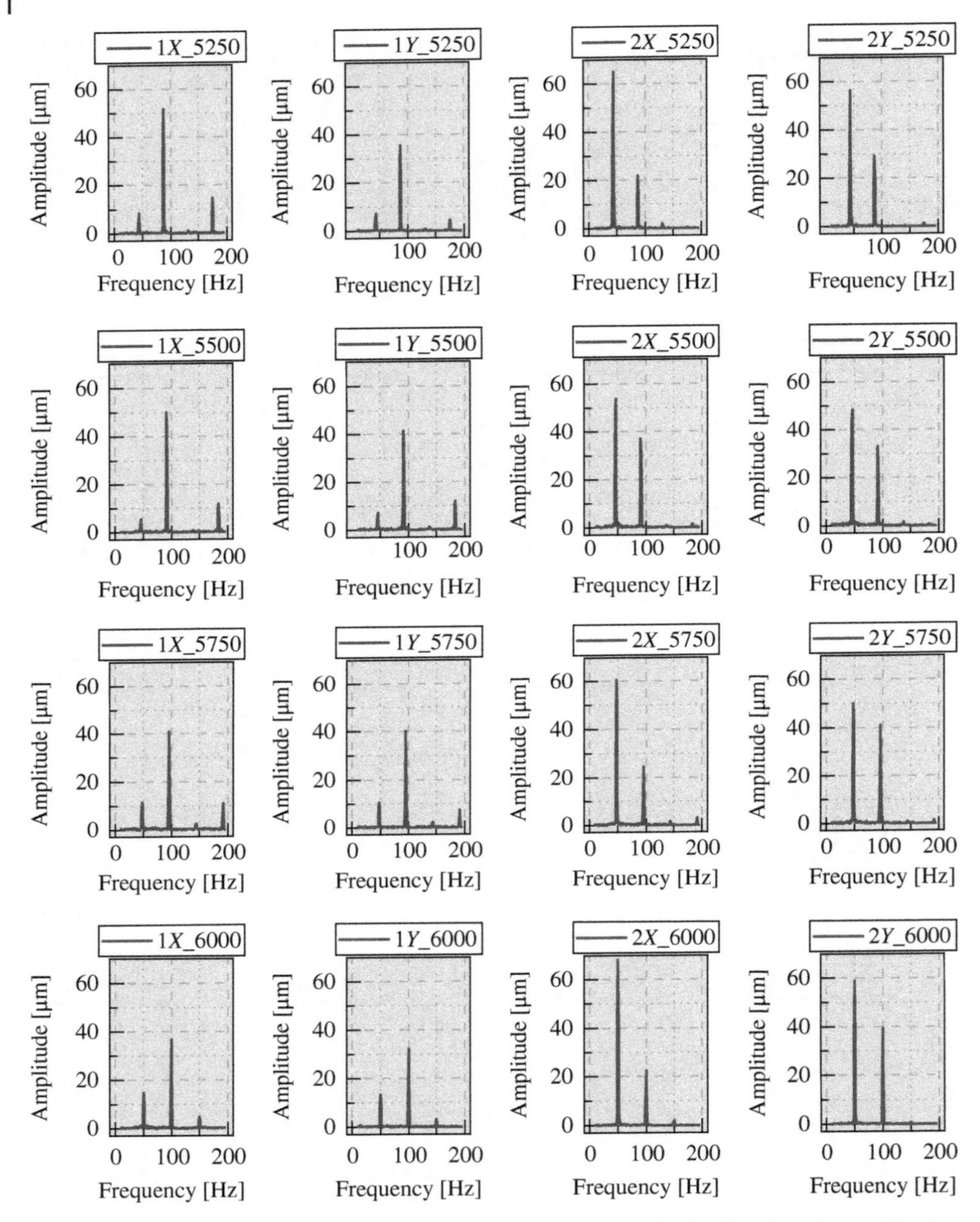

Figure A.4 FFT diagrams of the first and second bearings in the *X* and *Y* directions for rotational speeds of 5250–6000 rpm.

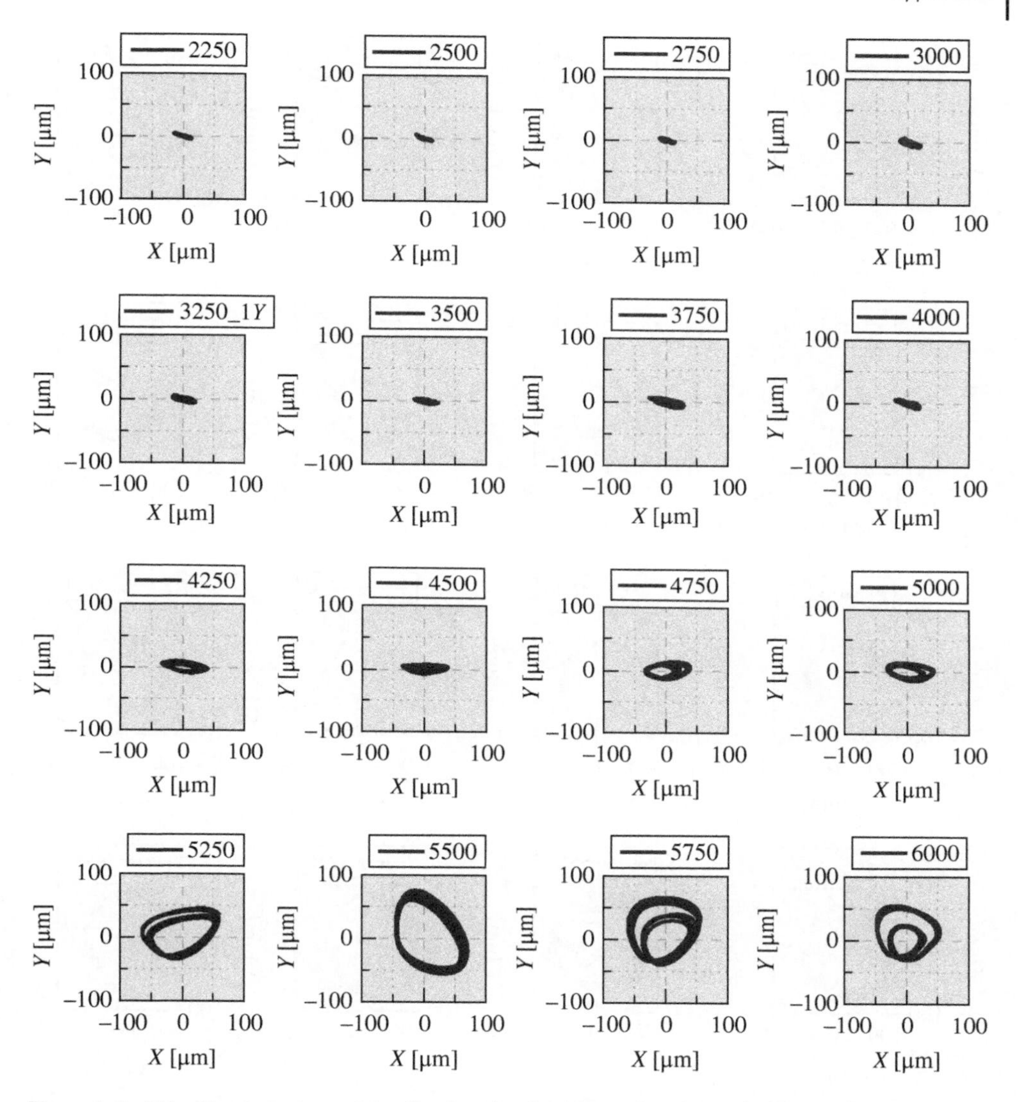

Figure A.5 Vibration trajectory of the first bearing for 16 rotational speeds, from 2250 to 6000 rpm.

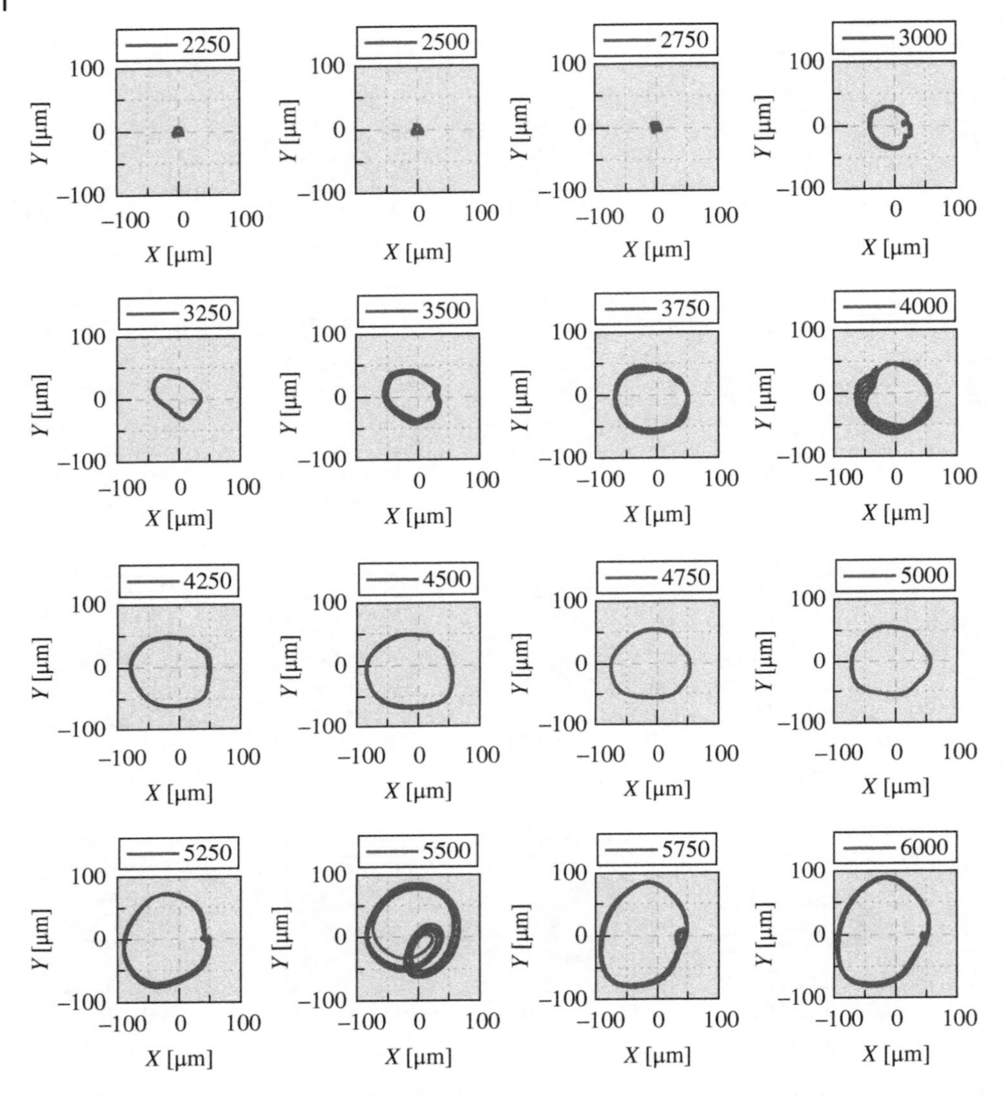

Figure A.6 Vibration trajectory of the second bearing for 16 rotational speeds, from 2250 to 6000 rpm.

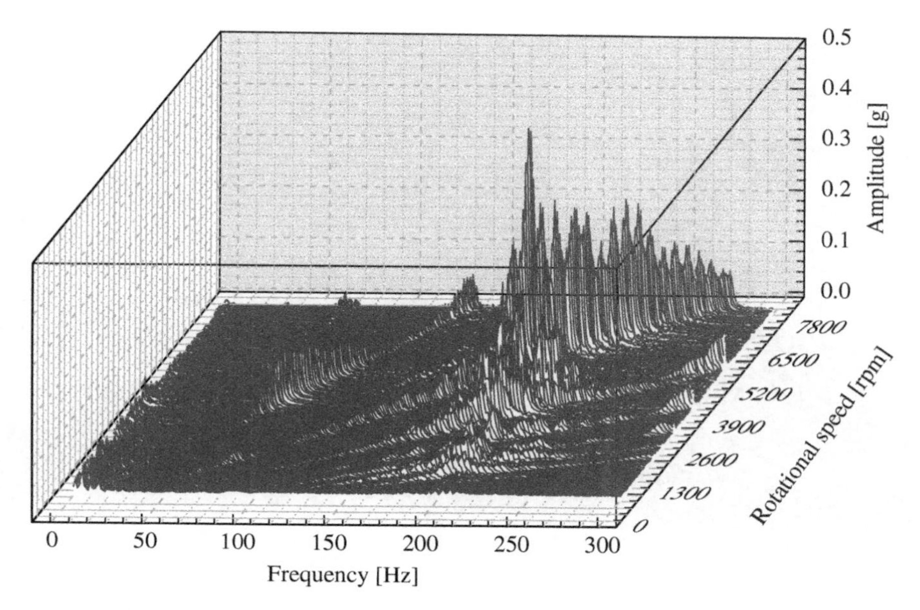

Figure A.7 Cascade diagram of an accelerometer placed on support no. 1 in the *X* direction (horizontal).

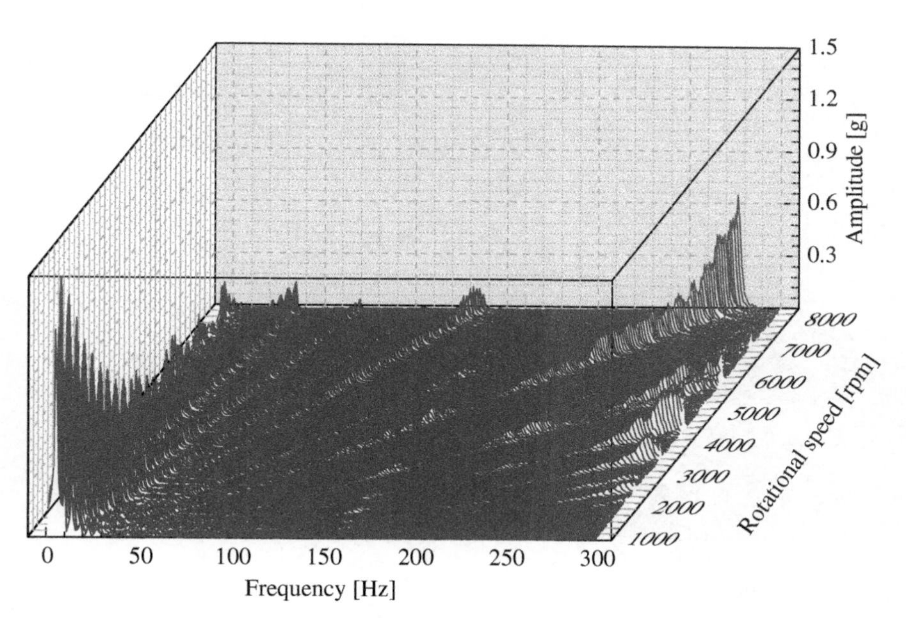

Figure A.8 Cascade diagram of an accelerometer placed on support no. 1 in the *Y* direction.

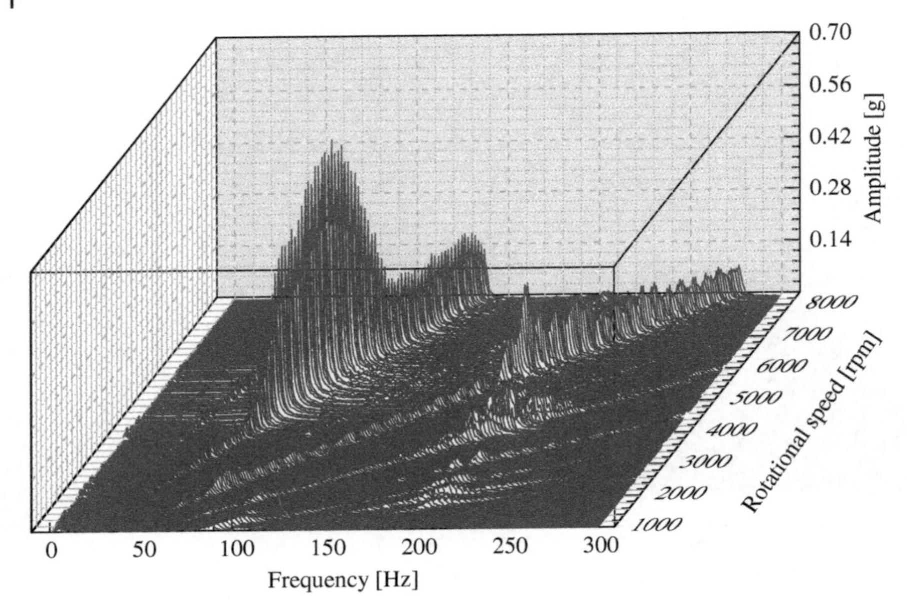

Figure A.9 Cascade diagram of an accelerometer placed on support no. 2 in the *X* direction.

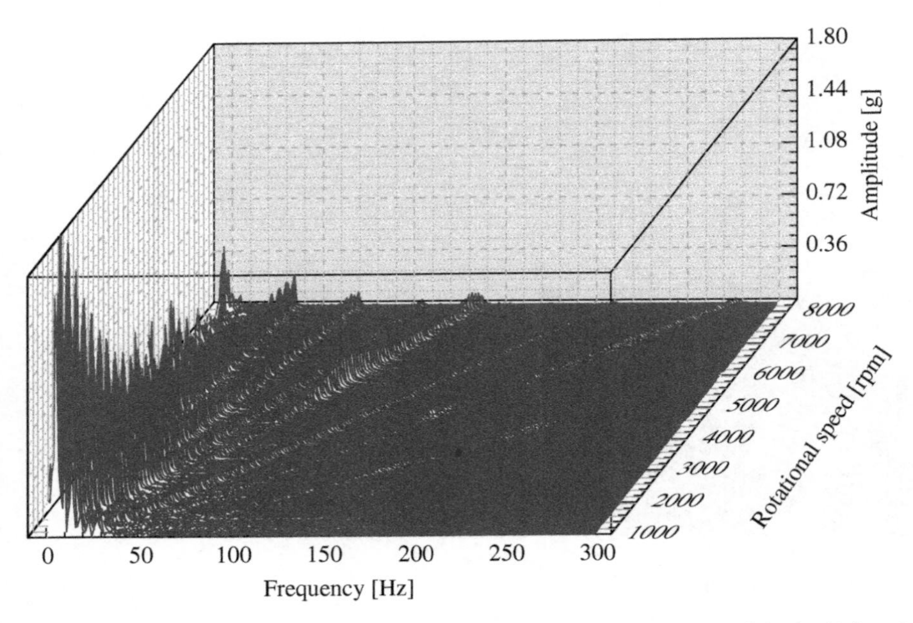

Figure A.10 Cascade diagram of an accelerometer placed on support no. 2 in the *Y* direction.

A.4 Rotor Axial Alignment Report

Rotor alignment is one of the operations carried out during the preparations of the laboratory test rig. The alignment process is described in Section 8.3. The results presented in Figures A.11 and A.12 prove compliance with the ISO 1940-1:2003 (2003) standard, which specifies the vibration requirements for rotor alignment.

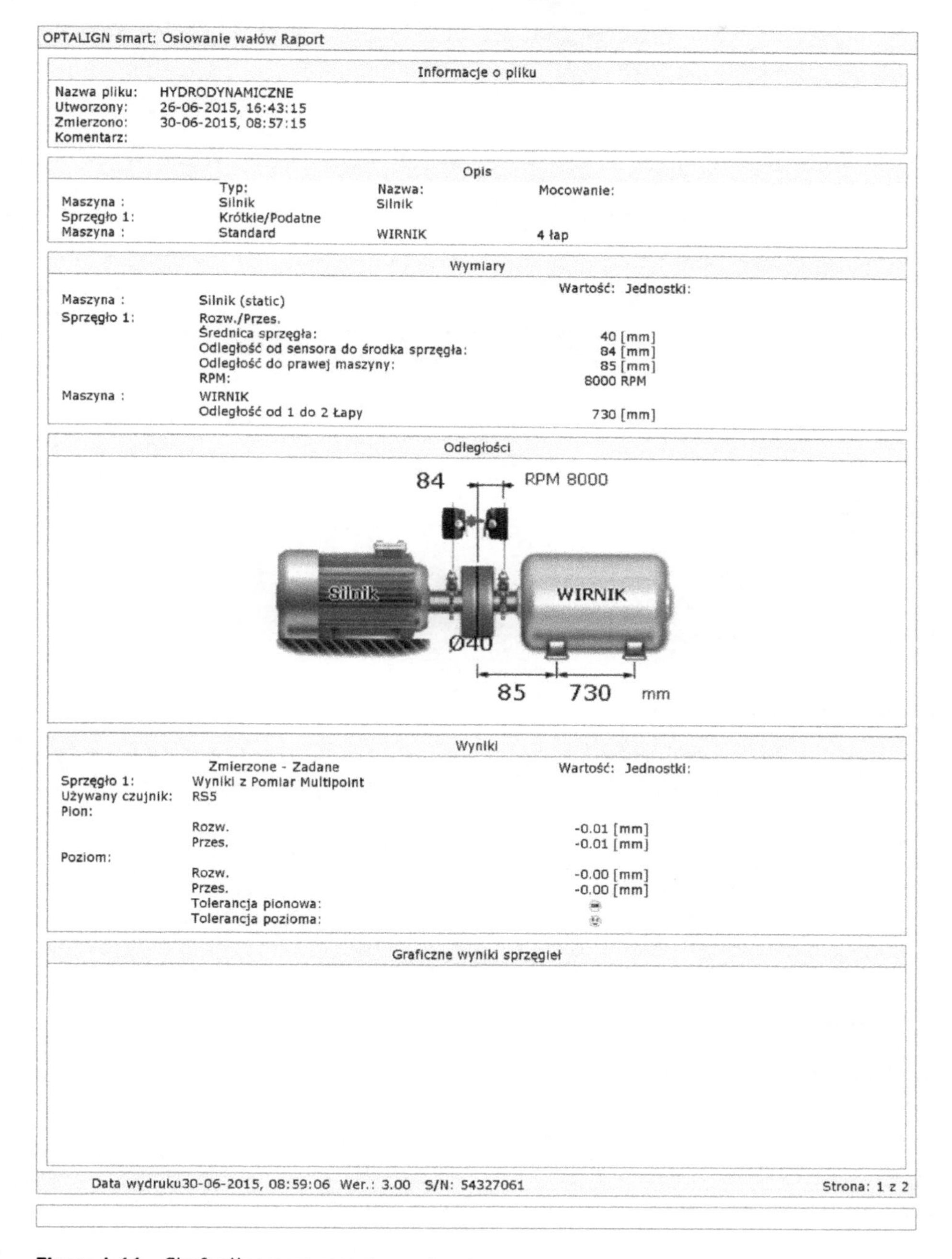

OPTALIGN smart: Osiowanie wałów Raport

Informacje o pliku

Nazwa pliku: HYDRODYNAMICZNE
Utworzony: 26-06-2015, 16:43:15
Zmierzono: 30-06-2015, 08:57:15
Komentarz:

Opis

	Typ:	Nazwa:	Mocowanie:
Maszyna :	Silnik	Silnik	
Sprzęgło 1:	Krótkie/Podatne		
Maszyna :	Standard	WIRNIK	4 łap

Wymiary

		Wartość:	Jednostki:
Maszyna :	Silnik (static)		
Sprzęgło 1:	Rozw./Przes.		
	Średnica sprzęgła:	40	[mm]
	Odległość od sensora do środka sprzęgła:	84	[mm]
	Odległość do prawej maszyny:	85	[mm]
	RPM:	8000	RPM
Maszyna :	WIRNIK		
	Odległość od 1 do 2 Łapy	730	[mm]

Odległości

Wyniki

	Zmierzone - Zadane	Wartość:	Jednostki:
Sprzęgło 1:	Wyniki z Pomiar Multipoint		
Używany czujnik:	RS5		
Pion:			
	Rozw.	-0.01	[mm]
	Przes.	-0.01	[mm]
Poziom:			
	Rozw.	-0.00	[mm]
	Przes.	-0.00	[mm]
	Tolerancja pionowa:		
	Tolerancja pozioma:		

Graficzne wyniki sprzęgieł

Data wydruku30-06-2015, 08:59:06 Wer.: 3.00 S/N: 54327061 Strona: 1 z 2

Figure A.11 Shaft alignment report – part one.

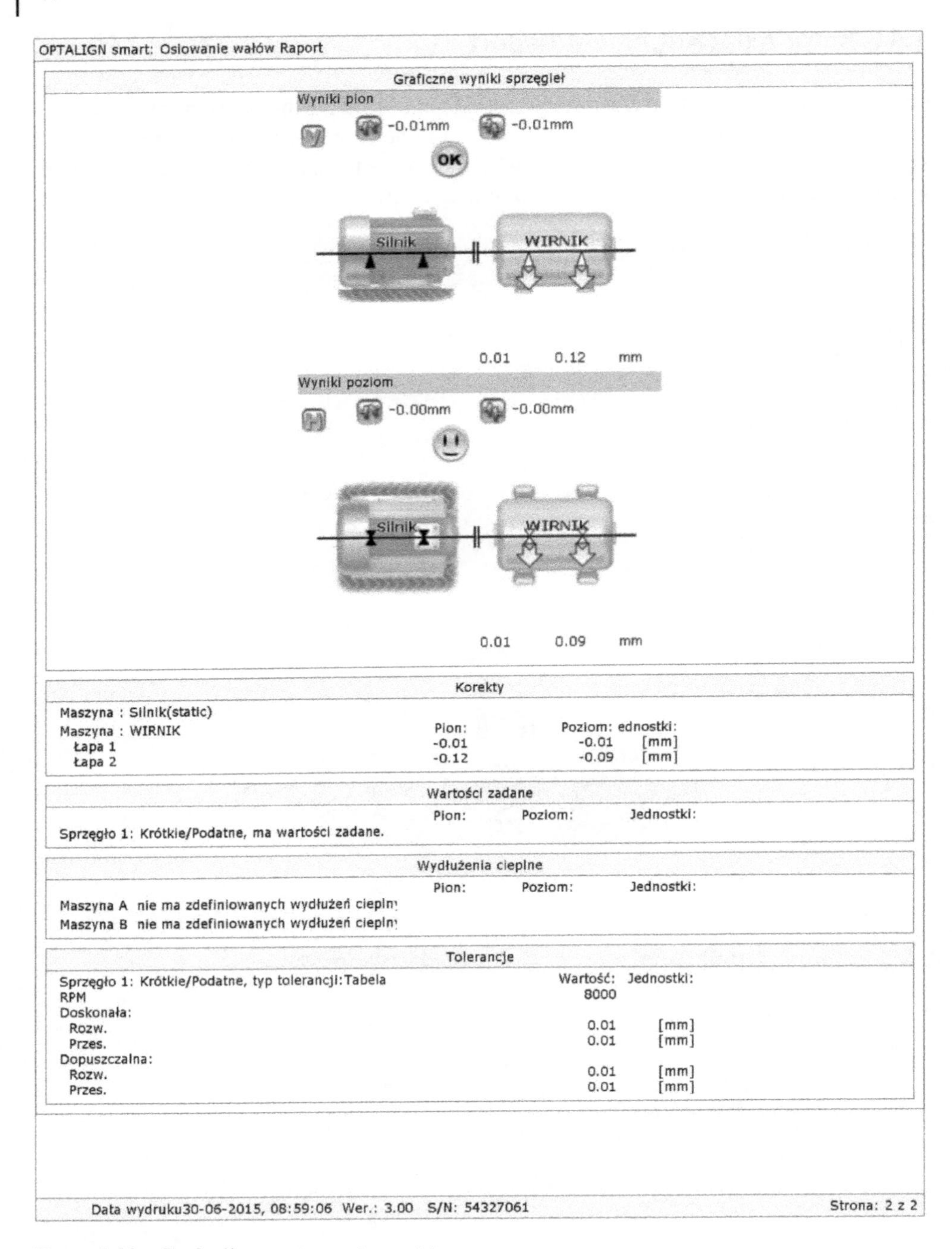

OPTALIGN smart: Osiowanie wałów Raport

Graficzne wyniki sprzęgieł

Wyniki pion

-0.01mm -0.01mm

OK

Wyniki poziom

-0.00mm -0.00mm

Korekty

Maszyna : Silnik(static)

Maszyna : WIRNIK	Pion:	Poziom:	ednostki:
Łapa 1	-0.01	-0.01	[mm]
Łapa 2	-0.12	-0.09	[mm]

Wartości zadane

	Pion:	Poziom:	Jednostki:
Sprzęgło 1: Krótkie/Podatne, ma wartości zadane.			

Wydłużenia cieplne

	Pion:	Poziom:	Jednostki:
Maszyna A nie ma zdefiniowanych wydłużeń ciepln			
Maszyna B nie ma zdefiniowanych wydłużeń ciepln			

Tolerancje

Sprzęgło 1: Krótkie/Podatne, typ tolerancji:Tabela	Wartość:	Jednostki:
RPM	8000	
Doskonała:		
Rozw.	0.01	[mm]
Przes.	0.01	[mm]
Dopuszczalna:		
Rozw.	0.01	[mm]
Przes.	0.01	[mm]

Data wydruku30-06-2015, 08:59:06 Wer.: 3.00 S/N: 54327061 Strona: 2 z 2

Figure A.12 Shaft alignment report – part two.

Appendix B

B.1 Fragments of Code of the Program Used for Processing the Signal from Experimental Tests

The interface of the program and its basic description is presented in Section 8.1. The "Signal" program is used to load experimental data, and then a signal after excitation and a reference signal are selected. The difference of these signals is saved as input data to the next program.

```
  % Code used for FFT analysis in the Matlab program Data_
length = length(dataX); % vector length of input samples
  w = hanning(Data_length); % defining the Hanning win-
dow dataX_w = w'.*dataX; % multiplication of data by the
Hanning window
  NFFT = 2 ^ nextpow2(Data_length); % vector length after the
FFT operation
FFTX = fft(dataX_in,NFFT); % Execution of the FFT
% declaration of filter parameters
  speed = 3250.0; % [rpm] rotational speed filterstar3250 =
((speed / 60.0)/3.0)-5; % [Hz] Filter start
  filterend3250 = ((speed / 60.0)*3.0)+5; % [Hz] Filter end
% band-pass filter
  d3250 = designfilt('bandpassiir','FilterOrder',20, 'Ha
lfPowerFrequency1',filterstar3250,'HalfPowerFrequenc
y2',filterend3250, 'SampleRate',51200); % filter design
% signal filtering
filterstable3250_1X = filtfilt(d3250,stable3250_1X); % filter-
ing operation
```

Bearing Dynamic Coefficients in Rotordynamics: Computation Methods and Practical Applications,
First Edition. Łukasz Breńkacz.

Companion website: www.wiley.com/go/brenkacz/bearingdynamiccoefficients

```
  % Deduction from the signal after excitation of the
reference signal Xaxis_value3 = floor(line_position_
displacement_X):floor(line_position_displacement_X_end); %
defining the data range data1XX_all = (data1XX(floor(line_
position_displacement_X):floor(line_position_displacement_X_
end)))) - (data1XX(floor(line_position_displacement_X_
begin):floor(line_position_displacement_X_end))); % Signal
subtraction operation in the time domain
  data2XX_all = (data2XX(floor(line_position_
displacement_X):floor(line_position_displacement_X_end)))) -
(data2XX(floor(line_position_displacement_X_begin):floor(line_
position_displacement_X_end))); % Signal subtraction operation
in the time domain
data1YX_all = (data1YX(floor(line_position_
displacement_X):floor(line_position_displacement_X_end))) -
(data1YX(floor(line_position_displacement_X_begin):floor(line_
position_
  displacement_X_end))); % Signal subtraction operation in the
time domain
data2YX_all = (data2YX(floor(line_position_
displacement_X):floor(line_position_displacement_X_end))) -
(data2YX(floor(line_position_displacement_X_begin):floor(line_
position_displacement_X_end))); % Signal subtraction operation
in the time domain

% Drawing a diagram
set(gcf, 'units', 'centimeters', 'pos', [5 5 figure_width
figure_height])
movegui(hFig, 'center'); % Moving the diagram to the center of
the screen
set(gcf, 'Color', 'w'); % Setting the background color
set(gca, 'Color', 'w'); % Setting the background color for the
coordinate system axes
set(gcf, 'Renderer', 'painters'); % Printer settings
hTitle    = title ('Force'); % Diagram title
hXLabel = xlabel('Time [s]'); % description of horizontal axis
of the diagram hYLabel = ylabel('Force [N]');

% description of vertical axis of the diagram
hLegend = legend('Force'); % Signature of the legend

% Setting the size and type of the font used on the diagram
set( gca                       , ...
    'FontName'   , FontName );
```

```
set([hTitle, hXLabel, hYLabel], ...
    'FontName'   , FontName);
 set([hLegend, gca]              , ...
     'FontSize'   , FontSize - 2);
set([hXLabel, hYLabel]  , ...
    'FontSize'   , FontSize     );
set( hTitle                     , ...
    'FontSize'   , FontSize     , ...
    'FontWeight' , 'bold'       );
grid on

% Drawing a diagram in the GUI after selecting the checkbox
if checkbox_value1XX
plot(Xaxis_value, data1XX, 'b');
%Drawing the start line of the part with excitation
  plot([line_position_displacement_X line_position_
displacement_X], [min(data1XX) max(data1XX)], '-','LineWidth',
2,'color','black');
  %Drawing the end line of the part with excitationplot([line_
position_displacement_X_end line_position_displacement_X_
end], [min(data1XX) max(data1XX)], '-','LineWidth',
2,'color','black');
  %Drawing the start line of the diagram with the reference
signal plot([line_position_displacement_X_begin line_position_
displacement_X_begin], [min(data1XX) max(data1XX)], '-.','Line-
Width', 2,'color','black');
%Drawing the end line of a diagram with the reference signal
plot([line_position_displacement_X_end line_position_
displacement_X_end], [min(data1XX) max(data1XX)], '-.','Line-
Width', 2,'color','black'); end
```

B.2 Fragments of Code of the Program Used for Calculation of Dynamic Coefficients of Bearings on the Basis of Experimental Tests

The "Calculation" program and its interface are presented in Section 8.2. Data prepared in the previously described "Signal" program can be loaded into the program. The program performs preliminary signal processing described in Section 8.5 and sets parameters for the analysis of dynamic coefficients of bearings. Due to the graphical user interface it is possible to display signals at different stages of calculation and to display dynamic coefficients of bearings.

Fragments of code of the program used for calculation of dynamic coefficients of bearings on the basis of experimental tests:

```
% Rotation of the coordinate system alpha = -pi()/4;
% rotation angle
% Formulae defining the transformation of the system
sxx1 =  data1XX*cos(alpha)+data1YX*sin(alpha);
syx1 = -data1XX*sin(alpha)+data1YX*cos(alpha);
sxy1 =  data1XY*cos(alpha)+data1YY*sin(alpha);
syy1 = -data1XY*sin(alpha)+data1YY*cos(alpha);
sxx2 =  data2XX*cos(alpha)+data2YX*sin(alpha);
syx2 = -data2XX*sin(alpha)+data2YX*cos(alpha);
sxy2 =  data2XY*cos(alpha)+data2YY*sin(alpha);
syy2 = -data2XY*sin(alpha)+data2YY*cos(alpha);
% Assignment of new displacement values (after coordinate sys-
tem rotation)
data1XX = -sxx1;
data2XX = -sxx2;
data1YY = -syy1;
data2YY = -syy2;
data1XY = -sxy1;
data2XY = -sxy2;
data1YX = -syx1;
data2YX = -syx2;

  % The Fast Fourier Transform Operation (FFT)
SampleVector=length(dataHX); % length of sample vector
  NFFT = 2^nextpow2(SampleVector); % length of sample vector
% execution of the FFT operation
FFTHX = fft(dataHX,NFFT)*MultiplierFFTH/SampleVector; FFT1HX =
fft(data1HX,NFFT)*MultiplierFFTH/SampleVector; FFT1HY;

  %% Calculation of stiffness and flexibility for the
first bearing
% Determination of impedance step 1
CCC1 = 0; % Zeroing of variables before calculation CCCTime1 =
0; %Zeroing before calculation DDD1 = 0; %Zeroing before calcu-
lation DDDTime1 = 0; %Zeroing before calculation

counterFFT = 1;
CCC1 = [FFT1HX(counterFFT), 0;
        0, FFT1HY(counterFFT)] * inv([FFT1XX(counterFFT),
        FFT1XY(counterFFT); FFT1YX(counterFFT),
        FFT1YYY(counterFFT)]);
```

```
for counterFFT = 2:1:Max;
CCCWhile1 = [FFT1HX(counterFFT), 0;
                 0, FFT1HY(counterFFT)] *
      inv([FFT1XX(counterFFT), FFT1XY(counterFFT);
                 FFT1YX(counterFFT), FFT1YY(counterFFT)]);

CCC1 = [CCC1; CCCWhile1];
end
% Transmitting the value to the main workspace
assignin('base','CCC1',CCC1)

% Determination of impedance step 2 - determination of
vectors H
counterH_start = 1;
counterH_end = counterH_start+1;
H1xx = CCC1(counterH_start, 1);
H1xy = CCC1(counterH_start, 2);
H1yx = CCC1(counterH_end, 1);
H1yy = CCC1(counterH_end, 2);

for AcounterDDD = 2:1:Max;
        counterH_start = counterH_start+2;
        counterH_end = counterH_end+2;

        H1xxWhile = CCC1(counterH_start, 1);
        H1xyWhile = CCC1(counterH_start, 2);
        H1yxWhile = CCC1(counterH_end, 1);
        H1yyWhile = CCC1(counterH_end, 2);

        H1xx = [H1xx; H1xxWhile];
        H1xy = [H1xy; H1xyWhile];
        H1yx = [H1yx; H1yxWhile];
        H1yy = [H1yy; H1yyWhile];

end

% Calculation of dynamic flexibility 1 step
counterDDD_start = 1;
counterDDD_end = counterDDD_start+1;

DDD1 = (CCC1(counterDDD_start:counterDDD_end,1:2))^-1;

for counterDDD = 2:1:Max;
    counterDDD_start = counterDDD_start+2;
    counterDDD_end = counterDDD_end+2;
```

```
DDDWhile1 = (CCC1(counterDDD_start:counterDDD_end,1:2))^-1;
DDD1 = [DDD1; DDDWhile1];
end

% Calculation of dynamic flexibility 2 step
counterF_start = 1;
counterF_end = counterF_start+1;
F1xx = DDD1(counterF_start, 1);
F1xy = DDD1(counterF_start, 2);
F1yx = DDD1(counterF_end, 1);
F1yy = DDD1(counterF_end, 2);
for counterDDD = 2:1:Max;
        counterF_start = counterF_start+2;
        counterF_end = counterF_end+2;
        F1xxWhile = DDD1(counterF_start, 1);
        F1xyWhile = DDD1(counterF_start, 2);
        F1yxWhile = DDD1(counterF_end, 1);
        F1yyWhile = DDD1(counterF_end, 2);

        F1xx = [F1xx;F1xxWhile];
        F1xy = [F1xy;F1xyWhile];
        F1yx = [F1yx;F1yxWhile];
        F1yy = [F1yy;F1yyWhile];

end
% Matrix I formation
i = sqrt(-1);
I = [];
temporaryI = [];
for counterI = Min:1:Max;
  temporaryI = [eye(2); eye(2); zeros(2); zeros(2)];
  I = [I;temporaryI];
End

% Matrix A formation
A =  [];
temporaryA = [];
for elA = Min:1:Max;
temporaryA = [real(DDD1(elA*2-1:elA*2,:)*[1, 0, 0, 0,
-(omega(elA))^2, 0, 0, 0, omega(elA)*i, 0, 0, 0;
                                          0, 1, 0, 0, 0,
-(omega(elA))^2, 0, 0, 0, omega(elA)*i, 0, 0]);
               real(DDD2(elA*2-1:elA*2,:)*[0, 0, 1, 0, 0, 0,
-(omega(elA))^2, 0, 0, 0, omega(elA)*i, 0;
                                          0, 0, 0, 1, 0, 0, 0,
```

```
-(omega(elA))^2, 0, 0, 0, omega(elA)*i]);
              imag(DDD1(elA*2-1:elA*2,:)*[1, 0, 0, 0,
-(omega(elA))^2, 0, 0, 0, omega(elA)*i, 0, 0, 0;
                                        0, 1, 0, 0, 0,
-(omega(elA))^2, 0, 0, 0, omega(elA)*i, 0, 0]);
              imag(DDD2(elA*2-1:elA*2,:)*[0, 0, 1, 0, 0, 0,
-(omega(elA))^2, 0, 0, 0, omega(elA)*i, 0;
                                        0, 0, 0, 1, 0, 0, 0,
-(omega(elA))^2, 0, 0, 0, omega(elA)*i]))];
    A = [A;temporaryA];
end

% Calculation of bearings dynamic coefficients
Z = A\I
```

Appendix C

This appendix presents the results of numerical calculations from the KINWIR, LDW and NLDW programs.

C.1 Journal Vibration Trajectories Calculated in KINWIR and LDW Software

The method of creating the diagrams is described in Chapter 8. Figure C.1 presents16 trajectory diagrams generated for the second bearing journal for rotational speeds from 2000 to 6000 rpm. Each diagram presents 12 revolutions of the rotor.

C.2 Journal Vibration Trajectories Calculated in NLDW Software

The method of creating the diagrams is described in Chapter 8. Figure C.2 presents16 trajectory diagrams generated for the second bearing journal for rotational speeds from 2000 to 6000 rpm. Each diagram represents 12 revolutions of the rotor.

Bearing Dynamic Coefficients in Rotordynamics: Computation Methods and Practical Applications,
First Edition. Łukasz Breńkacz.

Companion website: www.wiley.com/go/brenkacz/bearingdynamiccoefficients

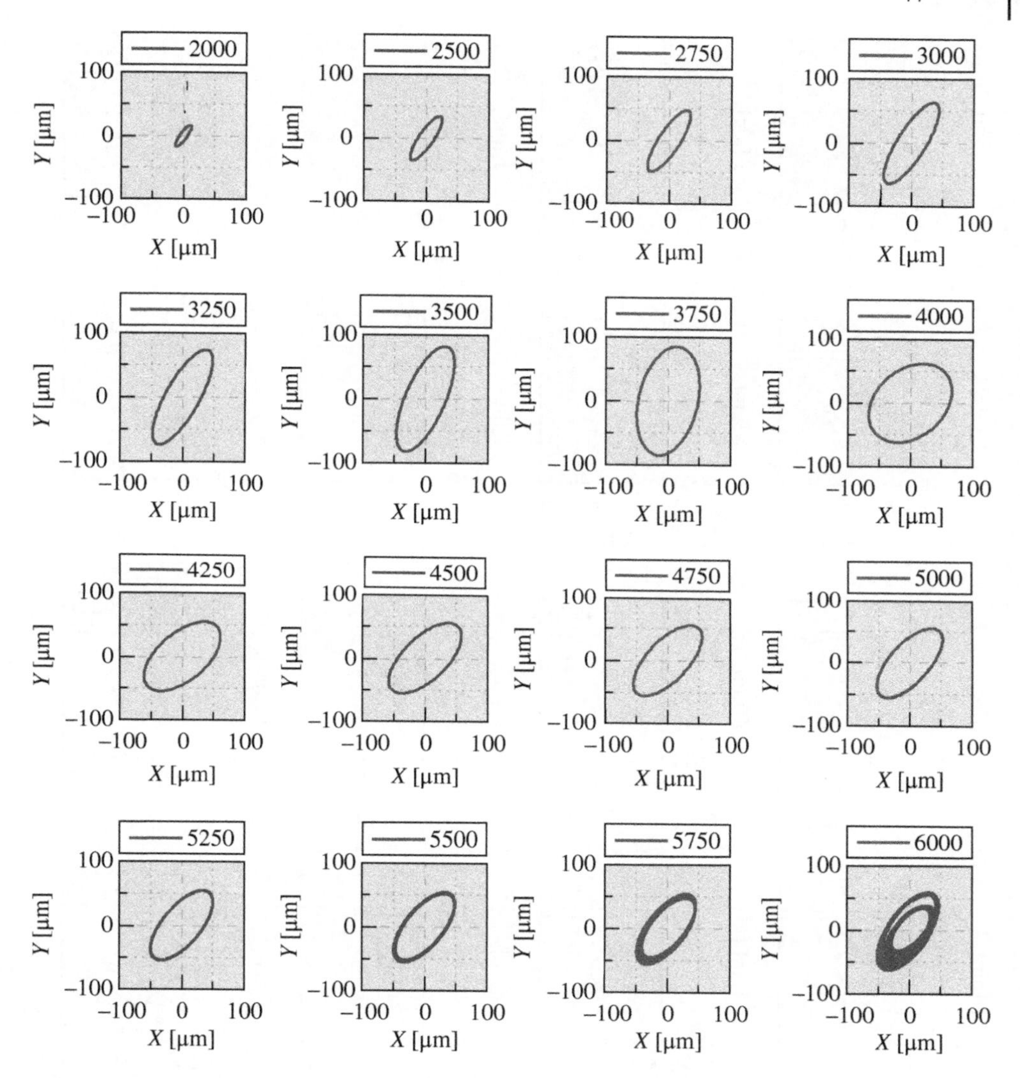

Figure C.1 Vibration trajectories calculated for the second bearing using a linear algorithm based on the KINWIR and LDW programs.

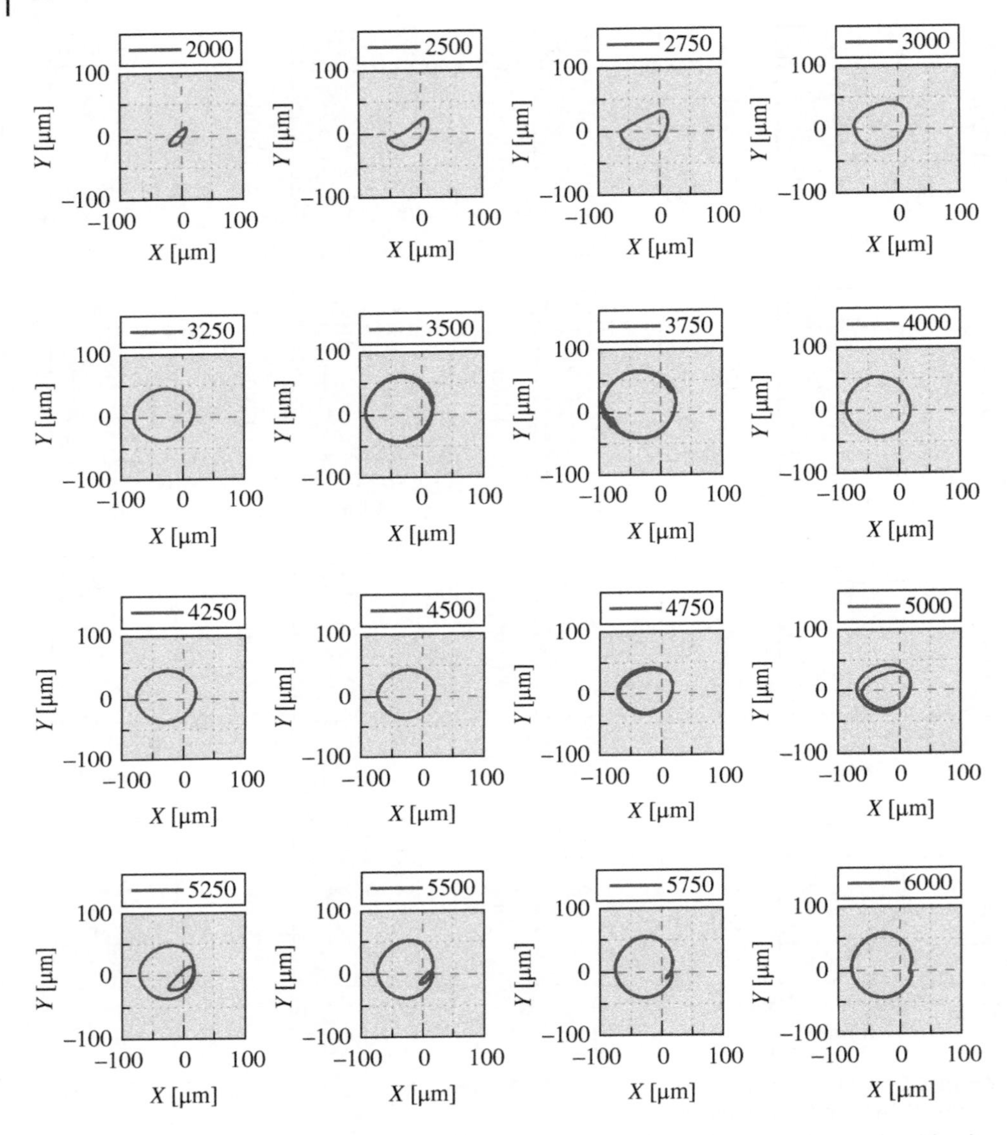

Figure C.2 Vibration trajectories calculated for the second bearing using a non-linear algorithm in the NLDW program.

Research Funding

- Some of the work was created within the framework of the ACTIVERING project entitled "Aktywne łożyska foliowe ze zmiennymi właściwościami dynamicznymi" (Active foil bearings with variable dynamic properties), agreement no. LIDER/51/0200/L9/17/NCBR/2018, implemented as part of the research and development program called "LIDER" (LEADER).
- The work was partly financed by the National Science Centre (Poland), with funds allocated by Decision No. 2015/17/N/ST8/01825 as part of the of the Preludium project.
- Some of the research was carried out as part of a one-year internship (from 1 April 2013 to 31 March 2014) at LMS International in Leuven, Belgium in the STA-DY-WI-CO project. The subject of the research task was: Identification of the nonlinear systems structural dynamics based on the rotating machinery test stand.
- Some of the work was carried out with the support of the key project POIG.01.01.01.02-00-016/08 – Model agroenergy complexes as an example of distributed cogeneration based on local renewable energy sources.
- Some of the work was carried out with the support of the NCBIR Strategic Programme: Advanced Technologies for Energy Generation. Research task no. 4 of the Strategic Programme: Development of integrated technologies in the production of fuel and energy from biomass, agricultural and other waste.
- Calculations were carried out at the Academic Computer Center in Gdansk.

Bearing Dynamic Coefficients in Rotordynamics: Computation Methods and Practical Applications, First Edition. Łukasz Breńkacz.

Companion website: www.wiley.com/go/brenkacz/bearingdynamiccoefficients

References

Abaqus (2015). Abaqus Analysis User Manual. Version 6.14-2.

Andreau, C., Ferdi, F., Ville, R., and Fillon, M. (2007). A method for determination of elastohydrodynamic behaviour of line shafting bearings in their environment. In: *Proceedings of the STLE/ASME International Joint Tribology Conference*, 153–155. San Diego, CA, USA (22–24 October 2007): ASME.

Arora, V., Van Der Hoogt, P.J.M., Aarts, R.G.K.M., and De Boer, A. (2011). Identification of stiffness and damping characteristics of axial air-foil bearings. *Int. J. Mech. Mater. Des.* 7: 231–243. https://doi.org/10.1007/s10999-011-9161-7.

Bannik, W.P. and Słuczajew, M.A. (1956). *Montaż turbin parowych*. Warszaw: Państwowe Wydawnictwa Techniczne.

Barwell, F.T. (1984). *Łożyskowanie*. Warszaw: Wydawnictwa Naukowo-Techniczne.

Batko, W. and Przysucha, B. (2014). Statistical analysis of the equivalent noise level. *Arch. Acoust.* 39: 195–198.

Batko, W., Dąbrowski, Z., and Kiciński, J. (2008). *Zjawiska nieliniowe w diagnostyce wibroakustycznej*. Wydawnictwo Naukowe Instytutu Technologii Eksploatacji - BIP.

Bielski, J. (2013). *Inżynierskie zastosowania systemu MES*. Krakow: Wydawnictwo PK.

Bishop, R.E.D., Gladwell, G.M.L., and Michaelson, S. (1972). *Macierzowa analiza drgań*. Warszaw: Wydawnictwa Naukowo-Techniczne.

Blok, H. (1975). Full journal bearings under dynamic duty: impulse method of solution and flapping action. *J. Lubr. Technol.* 97: 168–179.

Bonet, J. and Wood, R.D. (2009). *Nonlinear continuum mechanics for finite element analysis*, 2e. New York: Cambridge University Press.

Booker, J.F. (1965). Dynamically loaded journal bearings: mobility method of solution. *J. Basic Eng* 87: 537–546.

Breńkacz, Ł. (2015a). Identification of stiffness, damping and mass coefficients of rotor-bearing system using impulse response method. *J. Vibroeng.* 17: 2272–2282.

Breńkacz, Ł. (2015b). Identification of bearing dynamic coefficients with consideration of shaft unbalance (originally in Polish): Identyfikacja współczynników dynamicznych łożysk z uwzględnieniem niewyważenia wału. *Mechanik* 7: 57–64. http://dx.doi.org/10.17814/mechanik.2015.7.214.

Bearing Dynamic Coefficients in Rotordynamics: Computation Methods and Practical Applications,
First Edition. Łukasz Breńkacz.

Companion website: www.wiley.com/go/brenkacz/bearingdynamiccoefficients

Breńkacz, Ł. (2016). Adequacy ranges of linear and nonlinear methods for determining the dynamic properties of the rotating machinery (originally in Polish): Przedziały adekwatności liniowych i nieliniowych metod określania właściwości dynamicznych maszyn wirnikowych. Dotoral dissertation. Institute of Fluid Flow Machinery, Polish Academy of Sciences.

Breńkacz, Ł. and Żywica, G. (2016a). The sensitivity analysis of the method for identification of bearing dynamic coefficients. In: *Dynamical Systems: Modelling: Łódź Poland* (ed. J. Awrejcewicz), 81–96. Springer https://doi.org/10.1007/978-3-319-42402-6_8.

Breńkacz, Ł. and Żywica, G. (2016b). Numerical estimation of linear and nonlinear stiffness and damping coefficients of journal hydrodynamic bearings (originally in Polish): Numeryczne wyznaczanie liniowych i nieliniowych współczynników sztywności i tłumienia poprzecznych łożysk hydrodynamicz. *Mec. Dent.*: 648–649. https://doi.org/10.17814/mechanik.2016.7.108.

Breńkacz, Ł. and Żywica, G. (2017). Comparison of experimentally and numerically determined dynamic coefficients of the hydrodynamic slide bearings operating in the nonlinear rotating system. In: *Proceedings of ASME Turbo Expo 2017: Turbomachinery Technical Conference and Exposition*, 1–12. Charlotte, NC, USA (26–30 June 2017): ASME https://doi.org/10.1115/GT2017-64251.

Breńkacz, Ł., Żywica, G., and Drosińska-Komor, M. (2017a). The experimental identification of the dynamic coefficients of two hydrodynamic journal bearings operating at constant rotational speed and under nonlinear conditions. *Polish Marit. Res.* 24: 108–115. https://doi.org/10.1515/pomr-2017-0142.

Breńkacz, Ł., Żywica, G., Drosińska-Komor, M., and Szewczuk-Krypa, N. (2017b). *Dynamical Systems in Applications, the Experimental Determination of Bearings Dynamic Coefficients in a Wide Range of Rotational Speeds, Taking into Account the Resonance and Hydrodynamic Instability*. 13–24. Springer https://doi.org/10.1007/978-3-319-96601-4.

Buchacz, A., Świder, J., and Żółkiewski, S. (2013). *Modelowanie i analiza dynamiczna podatnych układów w ruchu obrotowym z uwzględnieniem tłumienia*. Gliwice: Wydawnictwo Politechniki Śląskiej.

Cannon, R.H. (1973). *Dynamika układów fizycznych*. Warszaw: Wydawnictwa Naukowo-Techniczne.

Chan, S.H. and White, M.F. (1991). Experimental determination of dynamic characteristics of a full size gas turbine tilting-pad journal bearing by an impact test method. In: *Proceedings of ASME Modal Analysis, Modeling, Diagnostics, and Control: Analytical and Experimental*, vol. 38, 291–298. ASME.

Chasalevris, A. (2016). Finite length floating ring bearings: operational characteristics using analytical methods. *Tribiol. Int.* 94: 571–590. https://doi.org/10.1016/j.triboint.2015.10.016.

Chmielniak, T.J. (1997). *Maszyny przepływowe*. Gliwice: Wydawnictwo Politechniki Śląskiej.

Chodkiewicz, R. (1998). *Ćwiczenia projektowe z turbin cieplnych*. Warszaw: Wydawnictwa Naukowo-Techniczne.

Cholewa, W. (2014a). Conditional contradictions in diagnostic knowledge bases. *Zesz. Nauk. Ciepl. Masz. Przepływowe – Turbomach.* 145: 33–34.

Cholewa, W. (2014b). *Intuitionistic Notice Boards for Expert Systems, Man-Machine Interactions*, 3e. Berlin: Springer.

Cholewa, W. and Kiciński, J. (1997). *Diagnostyka techniczna: Odwrotne modele diagnostyczne*. Gliwice: Wydawnictwo Politechniki Śląskiej.

Christopherson, D.G. (1941). A new mathematical method for the solution of film lubrication problems. *Proc. Inst. Mech. Eng.* 146: 129–135.

Cryer, C.W. (1971). The method of Christopherson for solving free boundary problems for infinite journal bearings by means of finite differences. *Math. Comput.* 25: 435–435. https://doi.org/10.1090/S0025-5718-1971-0298961-7.

Dąbrowski, Z. (2013). *Wały maszynowe*. Warszaw: Wydawnictwo Naukowe PWN.

Dacko, M., Borkowski, W., Dobrociński, S. et al. (1994). *Metoda Elementów Skończonych w mechanice konstrukcji*. Warszaw: Arkady.

Daniel, G.B., Machado, T.H., and Cavalca, K.L. (2016). Investigation on the influence of the cavitation boundaries on the dynamic behavior of planar mechanical systems with hydrodynamic bearings. *Mech. Mach. Theory* 99: 19–36. https://doi.org/10.1016/j.mechmachtheory.2015.11.019.

Dassault Systèmes (2020). Abaqus Unified FEA. http://www.3ds.com/products-services/simulia/products/abaqus (accessed 28 September 2020).

Delgado, A. (2015). Experimental identification of dynamic force coefficients for a 110 mm compliantly damped hybrid gas bearing. *J. Eng. Gas Turbines Power* 137: 072502.

Dimond, T.W., Sheth, P.N., Allaire, P.E., and He, M. (2009). Identification methods and test results for tilting pad and fixed geometry journal bearing dynamic coefficients – a review. *Shock Vib.* 16: 13–43. https://doi.org/10.3233/SAV-2009-0452.

Duff, J.N.M. and Curreri, J.R. (1960). *Drgania w technice*. Warszaw: Państwowe Wydawnictwa Techniczne.

Dunkerley, S. (1894). On the whirling and vibration of shafts. *Philos. Trans. R. Soc. Lond. A* 185: 279–360. https://doi.org/10.1098/rsta.1979.0079.

Easy LED (2014). Halogen lamp Easy LED PRO. https://easyled.fi/en/products/luminaires/pro/ (accessed 3 November 2020).

European Commission/CORDIS (2019). STAtic DYnamic piezo-driven streamWIse vortex generators for active flow COntrol. https://cordis.europa.eu/project/id/251309 (accessed 7 November 2020).

Fertis, D.G. (2010). *Nonlinear Structural Engineering: With Unique Theories and Methods to Solve Effectively Complex Nonlinear Problems*. Berlin: Springer.

Giergiel, J. (1990). *Tłumienie drgań mechanicznych*. Warszaw: Państwowe Wydawnictwo Naukowe.

Global Sensor Technology (2012). Eddy current sensor, Sinocera, Model CWY-DO-501A. https://globalsensortech.com/media/CWY-DO-501.pdf (accessed 5 November 2020).

Grabarski, A. and Wróbel, I. (2008). *Wprowadzenie do metody elementów skończonych*. Warszaw: Oficyna Wydawnicza Politechniki Warszawskiej.

Hagg, A.C. and Warner, P.C. (1953). Oil whip of flexible rotors. *Trans. ASME* 75: 1399–1344.

Hamrock, B.J., Schmid, S.R., and Jacobson, B.O. (2004). *Fundamentals of Fluid Film Lubrication*, 2e. New York: Marcel Dekker.

Hayashi, C. (1964). *Drgania nieliniowe w układach fizycznych*. Warszaw: Wydawnictwa Naukowo-Techniczne.

Helwany, S. (2007). *Applied Solid Mechanics with ABAQUS Applications*. Wiley.

Illner, T., Bartel, D., and Deters, L. (2015). Determination of the transition speed in journal bearings under consideration of bearing deformation. *Tribol. Int.* 82: 58–67. https://doi.org/10.1016/j.triboint.2014.09.023.

ISO 10816-1:1995 (1995). *Mechanical vibration – Evaluation of machine vibration by measurements on non-rotating parts – Part 1:General guidelines.*

ISO 10817-1:1998 (1998). *Rotating shaft vibration measuring systems – Part 1: Relative and absolute sensing of radial vibration.*

ISO 1940-1:2003 (2003). *Mechanical vibration - balance quality requirements for rotors in a constant (rigid) state. Part 1: Specification and verification of balance tolerances.*

Jankowski, A. (1983). *Algorytmy metod numerycznych*, 3e. Gdansk: Wydawnictwo Politechniki Gdańskiej.

Jáuregui, J.C., Andrés, L.S., and De Santiago, O. (2012). Identification of bearing stiffness and damping coefficients using phase-plane diagrams. In: *Proceedings of the ASME Turbo Expo 2012 – Turbine Technical Conference and Exposition*, 731–737. Copenhagen, Denmark (11–15 June 2012): ASME.

Kazimierski, Z. and Krysiński, J. (1981). *Łożyskowanie gazowe i napędy mikroturbinowe.* Warszaw: Wydawnictwa Naukowo-Techniczne.

Kiciński, J. (1988). Teoretyczny model termoelastohydrodynamiczny poprzecznych łożysk ślizgowych. *Trybologia* 1: 4.

Kiciński, J. (1989). New method of description of dynamic properties of slide bearings. *Wear* 132: 205–220.

Kiciński, J. (1994). *Teoria i badania hydrodynamicznych poprzecznych łożysk ślizgowych.* Wrocław: Ossolineum.

Kiciński, J. (2005). *Dynamika wirników i łożysk ślizgowych.* Gdansk: Maszyny Przepływowe.

Kiciński, J. (2006). *Rotor Dynamics.* IFFM Publisher.

Kiciński, J. and Żywica, G. (2012). The numerical analysis of the steam microturbine rotor supported on foil bearings. *Adv. Vib. Eng.* 11: 113–119.

Kiciński, J. and Żywica, G. (2014a). *Steam Microturbines in Distributed Cogeneration.* Springer.

Kiciński, J. and Żywica, G. (2014b). *Dynamika mikroturbin parowych.* Gdansk: Wydawnictwa Instytutu Maszyn Przepływowych PAN.

Kiciński, J., Drozdowski, R., and Materny, P. (1997). The non-linear analysis of the effect of support construction properties on the dynamic properties of multi-support rotor systems. *J. Sound Vib.* 206: 523–539. https://doi.org/10.1006/jsvi.1997.1113.

Kłosowski, P. (2011). *Metody numeryczne w mechanice konstrukcji z przykładami w programie Matlab.* Gdansk: Wydawnictwo Politechniki Gdańskiej.

Komorska, I. (2011). *Vibroacoustic Diagnostic Model of the Vehicle Drive System.* Radom: Institute for Sustainable Technologies.

Kosmol, J. (2010). *Wybrane zagadnienia metodologii badań.* Gliwice: Wydawnictwo Politechniki Śląskiej.

Kotulski, Z. and Szczepiński, W. (2004). *Rachunek błędów dla inżynierów.* Warszaw: Wydawnictwa Naukowo-Techniczne.

Kozánek, J. and Půst, L. (2011). Spectral properties and identification of aerostatic bearings. *Acta Mech. Sin.* 27: 63–67. https://doi.org/10.1007/s10409-011-0407-2.

Kozánek, J., Simek, J., Steinbauer, P., and Bílkovský, A. (2009). Identification of stiffness and damping coefficients of aerostatic journal bearing. *Eng. Mech.* 16: 209–220.

Kruszewski, J. and Wittbrodt, E. (1992). *Drgania układów mechanicznych w ujęciu komputerowym. Tom I zagadnienia liniowe*. Warszaw: Wydawnictwa Naukowo-Techniczne.

Kruszewski, J., Wittbrodt, E., and Walczyk, Z. (1996). *Drgania układów mechanicznych w ujęciu komputerowym. Tom II zagadnienia wybrane*. Warszaw: Wydawnictwa Naukowo-Techniczne.

Krzyżanowski, P. (2011). *Obliczenia inżynierskie i naukowe. Szybkie, skuteczne, efektowne*. Wydawnictwo Naukowe PWN: Warszaw.

Kucharski, T. (2000). *Programowanie obliczeń inżynierskich*. Gdansk: Wydawnictwo Politechniki Gdańskiej.

Łączkowski, R. (1974). *Drgania elementów turbin cieplnych*. Warszaw: Wydawnictwa Naukowo-Techniczne.

Lent, S.C. (2013). *Learning to Program with MATLAB: Building GUI Tools*, 1e. Notre Dame, IN: University of Notre Dame.

Lipka, J. (1967). *Wytrzymałość maszyn wirnikowych*. Warszaw: Wydawnictwa Naukowo-Techniczne.

Lund, J.W. (1987). Review of the concept of dynamic coefficients for fluid film journal bearings. *J. Tribol.* 109: 37–41. https://doi.org/10.1115/1.3261324.

Lyons, R.G. (2010). *Wprowadzenie do cyfrowego przetwarzania sygnałów*. Warszaw: Wydawnictwa Komunikacji i Łączności.

Maia, N.M.M., Silva, J.M.M., He, J. et al. (1998). *Theoretical and Experimental Modal Analysis*. Wiley.

Majchrzak, E. and Mochnacki, B. (2004). *Metody numeryczne podstawy teoretyczne, aspekty praktyczne i algorytmy*. Gliwice: Wydawnictwo Politechniki Śląskiej.

MBJ Electronics (2011). Diamond 401A. Data sheet.

Meruane, V. and Pascual, R. (2008). Identification of nonlinear dynamic coefficients in plain journal bearings. *Tribol. Int.* 41: 743–754. https://doi.org/10.1016/j.triboint.2008.01.002.

Mikielewicz, J., Kiciński, J., Cholewa, W. et al. (2005). *Modelowanie i diagnostyka oddziaływań mechanicznych, aerodynamicznych i magnetycznych w turbozespołach energetycznych*. Gdansk: Wydawnictwo IMP PAN.

Milenin, A. (2010). *Podstawy Metody Elementów Skończonych. Zagadnienia termomechaniczne*. Krakow: Wydawnictwa AGH.

Miller, A. (1985). *Maszyny i urządzenia cieplne i energetyczne*. Warszaw: Wydawnictwa Szkolne i Pedagogiczne.

Miller, B.A. and Howard, S.A. (2009). Identifying bearing rotor-dynamic coefficients using an extended kalman filter. *Tribol. Trans.* 52: 671–679. https://doi.org/10.1080/10402000902913295.

Minorsky, N. (1967). *Drgania nieliniowe*. Warszaw: Państwowe Wydawnictwo Naukowe.

Mrozek, B. and Zbigniew, M. (2010). *MATLAB i Simulink. Poradnik użytkownika*, 3e. Gliwice: Helion.

Muszyńska, A. (2005). *Rotordynamics*. Boca Raton, FL: Taylor & Francis.

Neyman, A. and Sikora, J. (1999). *Hydrodynamiczne łożyska ślizgowe poprzeczne*. Gdansk: Wydawnictwo Politechniki Gdańskiej.

Niezgodziński, M.E. and Niezgodziński, T. (2004). *Wzory wykresy i tablice wytrzymałościowe*, 9e. Warszaw: Wydawnictwa Naukowo-Techniczne.

Nikiel, T. (1956). *Turbiny parowe*. Warszaw: Państwowe Wydawnictwa Techniczne.

Nordmann, R. and Schoellhorn, K. (1980). Identification of stiffness and damping coefficients of journal bearings by means of the impact method. In: *Proceedings of the 2nd International Conference on Vibrations in Rotating Machinery*, 231–238. Vienna: Springer.

Orłowski, W. and Słowański, L. (1978). *Wytrzymałość materiałów przykłady obliczeń*, 3e. Warszaw: Wydawnictwo Arkady.

PCB Piezotronics (2007). Impact hammer PCB Piezotronics ICP 086D05. https://www.pcb.com/products?model=086D05 (accessed 3 November 2020).

PCB Piezoelectronics (2008). Accelerometers calibrator PCB 699A02. https://www.pcb.com/products?m=699B02 (accessed 3 November 2020).

PCB Piezotronics (2011). Single-axis accelerometer 608A11. https://www.pcb.com/products?m=608A11 (accessed 3 November 2020).

Phantom (2015). High-speed camera Phantom v2512. https://www.phantomhighspeed.com/products/cameras/ultrahighspeed/v2512 (accessed 3 November 2020).

PN-93/N-01359 (1993). *Drgania mechaniczne. Wyważenie wirników sztywnych. Wyznaczanie dopuszczalnego niewyważenia resztkowego.*

Przysucha, B., Batko, W., and Szeląg, A. (2015). Analysis of the accuracy of uncertainty noise measurement. *Arch. Acoust.* 40: 183–189.

Qiu, Z.L. and Tieu, A.K. (1997). Identification of sixteen force coefficients of two journal bearings from impulse responses. *Wear* 212: 206–212. https://doi.org/10.1016/S0043-1648(97)00154-3.

Radwańska, M. (2010). *Metody komputerowe w wybranych zagadnieniach mechaniki konstrukcji*, 3e. Krakow: Wydawnictwo PK.

Rakowski, G. and Kacprzyk, Z. (1993). *Metoda Elementów Skończonych w Mechanice Konstrukcji*. Warszaw: Oficyna Wydawnicza Politechniki Warszawskiej.

Reynolds, O. (1886). On the theory of lubrication and its applications to Mr. Beauchamp Tower's experiments, including an experimental determination of the viscosity of olive oil. *Proc. R. Soc.* 40: 191–203.

Rucka, M. and Krzysztof, W. (2012). *Dynamika budowli z przykładami w środowisku Matlab*, 2e. Gdansk: Wydawnictwo Politechniki Gdańskiej.

Sathyamoorthy, M. (2000). *Nonlinear Analysis of Structures*. Boca Raton, FL: CRC Press.

Sikora, J. (2009). *Problemy badawcze wytrzymałości zmęczeniowej warstw ślizgowych łożysk*. Gdansk: Wydawnictwo Politechniki Gdańskiej.

Skup, Z. (2010). *Zjawiska nieliniowe w tłumieniu drgań*. Warszaw: Oficyna Wydawnicza Politechniki Warszawskiej.

Smith, S.T. (2006). *MATLAB Advanced GUI Development*. Dog Ear Publishing.

Smolaga, K. (1959). *Obsługa turbin parowych*. Warszaw: Państwowe Wydawnictwa Techniczne.

Someya, T. (1989). *Journal-Bearing Databook, Climate Change 2013 – The Physical Science Basis*. Berlin: Springer https://doi.org/10.1007/978-3-642-52509-4.

Sommerfeld, A. (1904). Zur hydrodynamischen theorie der SchmiermitteIreiburg. *Zeitschr. Angew. Math. Phys.* 50: 97–155.

Stieber, W. (1933). *Das schwimmlager: Hydrodynamische theorie des gleitlagers*. VDI-Verlag GMbH.

Swift, H.W. (1932). The stability of lubricationg film in journal bearings. *Proc. Inst. Civil Eng.* 233: 267–288. https://doi.org/10.1680/imotp.1932.13239.

Szymczyk, E. (2006). *Matlab dla Mechaników*. Warszaw: Wojskowa Akademia Techniczna.

Tiwari, R. and Chakravarthy, V. (2009). Simultaneous estimation of the residual unbalance and bearing dynamic parameters from the experimental data in a rotor-bearing system. *Mech. Mach. Theory* 44: 792–812. https://doi.org/10.1016/j.mechmachtheory.2008.04.008.

Tiwari, R., Lees, A.W., and Friswell, M.I. (2004). Identification of dynamic bearing parameters: a review. *Shock Vib. Digest* 36: 99–124. https://doi.org/10.1177/0583102404040173.

Uhl, T. (1997). *Komputerowo wspomagana identyfikacja modeli konstrukcji mechanicznych*. Warszaw: Wydawnictwa Naukowo-Techniczne.

Uhl, T. (ed.) (2006). *Wybrane zagadnienia analizy modalnej konstrukcji mechanicznych*. Krakow: Akademia Górniczo-Hutnicza.

Uhl, T. (ed.) (2008). *Wybrane zagadnienia analizy modalnej konstrukcji mechanicznych*. Krakow: Akademia Górniczo-Hutnicza.

Uhl, T. and Lisowski, W. (1996). *Praktyczne problemy analizy modalnej konstrukcji*. Krakow: Wydawnictwo AGO.

Wang, Y.P. and Kim, D. (2013). Experimental identification of force coefficients of large hybrid air foil bearings. In: *Proceedings of the ASME Turbo Expo 2013 – Turbine Technical Conference and Exposition*. San Antonio, TX, USA (3–7 June 2013): ASME.V07BT30A026

Wasilczuk, M. (2012). *Wielkogabarytowe hydrodynamiczne łożyska wzdłużne*. Wydawnictwo Naukowe Instytutu Technologii Eksploatacji - BIP.

Wierzcholski, K. (1994). *Elementy obliczania łożysk ślizgowych*. Szczecin: Wydawnictwo Akademii Rolniczej w Szczecinie.

Wiśniewski, S. (1974). *Obciążenia cieplne silników turbinowych*. Warszaw: Wydawnictwa Komunikacji i Łączności.

Zapomel, J., Ferfecki, P., and Kozánek, J. (2014). Application of the Monte Carlo method for investigation of dynamical parameters of rotors supported by magnetorheological squeeze film damping devices. *Appl. Comput. Mech.* 8: 129–138.

Zhang, X., Yin, Z., Gao, G., and Li, Z. (2015). Determination of stiffness coefficients of hydrodynamic water-lubricated plain journal bearings. *Tribol. Int.* 85: 37–47. https://doi.org/10.1016/j.triboint.2014.12.019.

Zhang, Y.Y., Xie, Y.B., and Qiu, D.M. (1992a). Identification of linearized oil-film coefficients in a flexible rotor-bearing system, part II: experiment. *J. Sound Vib.* 2: 549–559.

Zhang, Y.Y., Xie, Y.B., and Qiu, D.M. (1992b). Identification of linearized oil-film coefficients in a flexible rotor-bearing system, part I: model and simulation. *J. Sound Vib.* 2: 531–547.

Zhou, J., Di, L., Cheng, C. et al. (2016). A rotor unbalance response based approach to the identification of the closed-loop stiffness and damping coefficients of active magnetic bearings. *Mech. Syst. Signal Process.* 66–67: 665–678. https://doi.org/10.1016/j.ymssp.2015.06.008.

Zieliński, Z. (1969). *Maszyny i urządzenia cieplne i energetyczne*. Warszaw: Państwowe Wydawnictwa Szkolenia Zawodowego.

Zienkiewicz, O.C. and Taylor, R.L. (2000). *The Finite Element Method: Volume 1: The Basis*, 689. Butterworth-Heinemann.

Żywica, G. (2008). Analysis of the supporting structure defects concerning the dynamic state of the rotating machine (originally in Polish): Analiza defektów konstrukcji podpierającej w odniesieniu do stanu dynamicznego maszyny wirnikowej. Doctoral dissertation. Institute of Fluid Flow Machinery, Polish Academy of Sciences.

Index

Bearing Dynamic Coefficients in Rotordynamics: Computation Methods and Practical Applications,
First Edition. Łukasz Breńkacz.

Companion website: www.wiley.com/go/brenkacz/bearingdynamiccoefficients